Introduction to Postharvest Management

Introduction to Postharvest Management

Scott Simmons

MURPHY & MOORE
www.murphy-moorepublishing.com

Published by Murphy & Moore Publishing,
1 Rockefeller Plaza,
New York City, NY 10020, USA
www.murphy-moorepublishing.com

Introduction to Postharvest Management
Scott Simmons

International Standard Book Number: 978-1-63987-335-7 (Hardback)

Cataloging-in-Publication Data

Introduction to postharvest management / Scott Simmons.
p. cm.
Includes bibliographical references and index.
ISBN 978-1-63987-335-7
1. Crops--Postharvest technology. 2. Food crops--Postharvest technology.
3. Postharvest diseases and injuries. 4. Agricultural processing. I. Simmons, Scott.

SB129 .I58 2022

631.55--dc23

Table of Contents

Preface

This book is a culmination of my many years of practice in this field. I attribute the success of this book to my support group. I would like to thank my parents who have showered me with unconditional love and support and my peers and professors for their constant guidance.

The process of gathering crops from the field is known as harvesting. The stage that immediately follows crop production is known as postharvest management. It includes cleaning, sorting, cooling and packing. It aims to maintain the quality of the crops after harvest. Some of the goals of this field are avoiding moisture loss and slowing down unwanted chemical changes, keeping the product cool and preventing physical damages to the crops. Many devices like conveyor belts, walk-in coolers and automated sorting and packing stations are used to maintain the final quality of the crop. Sophisticated refrigerated and atmosphere controlled environments, and high-speed cooling are some of the methods that used to prolong the freshness of the production. This book is a valuable compilation of topics, ranging from the basic to the most complex theories and principles in this field. It explores all the important aspects of postharvest management in the present day scenario. This book will serve as a reference to a broad spectrum of readers.

The details of chapters are provided below for a progressive learning:

Chapter - What is Postharvest Management?

The stage of crop production which falls immediately after the harvest period is known as postharvest. It includes the management of cleaning, packing and sorting of crops. This is an introductory chapter which will briefly explain the postharvest system, postharvest handling, postharvest physiology and the challenges associated with postharvest management.

Chapter - Postharvest Processing

Postharvest processing is a crucial stage of postharvest management. It helps in preventing spoilage of crops. Some of the processes of this domain are ultraviolet light treatment, pre-cooling, fruit waxing, drying, sorting and grading, etc. This chapter explores these processes to provide an extensive understanding of the subject.

Chapter - Postharvest Handling of Fruits and Vegetables

It is important to ensure the fruits and vegetables are handled with utmost care in the postharvest period. This chapter has been carefully written to provide an easy understanding of the various facets of postharvest handling of fruits and vegetables. Some of the fruits and vegetables focused in this chapter are apples, mangoes, tomatoes, beans among others.

Chapter - Postharvest Storage of Crops

Different crops require different temperatures and techniques to be stored after harvest in order to avoid any degradation in the quality. This chapter discusses in detail the theories and methodologies related to the postharvest storage of different crops such as grains, onions, etc.

Chapter - Postharvest Pathogens and Disease Management

Crops are prone to infectious pathogens after harvest. Such pathogens can also lead to diseases. Common disease causing pathogens include alternaria, sclerotinia, colletotrichum, etc. The topics elaborated in this chapter will help in getting a better perspective about pathogens, the diseases caused by them and their management strategies.

Chapter - Postharvest Packaging

Postharvest packaging is an important aspect of postharvest management. Temperature management is essential for packaging crops. All the diverse aspects of postharvest packaging such as packaging requirements, the different types of packaging and modified atmosphere packaging have been carefully analyzed in this chapter.

Scott Simmons

1

What is Postharvest Management?

The stage of crop production which falls immediately after the harvest period is known as postharvest. It includes the management of cleaning, packing and sorting of crops. This is an introductory chapter which will briefly explain the postharvest system, postharvest handling, postharvest physiology and the challenges associated with postharvest management.

Post-harvest management is a system of handling, storing, and transporting agricultural commodities after harvest. For some commodities such as coffee and cocoa, post-harvest activities may include drying and fermenting as well. The importance of post-harvest management has been established over the years—it strengthens the action chain that produces transports, and processes food and all other related products that give sustenance to the world population. Moreover, post-harvest management allows the effective planning of how to sustainably manage finite resources for the future. In other words, through effective post-harvest management, action can be done now so that the environment can sustain future generations.

During the post-harvest period, handlers and producers focus on preserving quality, quantity, and the safety of the commodities. For coffee and cocoa producers and traders, maintaining the integrity of the beans is important because it directly affects the price of the commodity. This means having to ensure that moisture, contaminants, and insects will not affect the quality of the commodities.

While extremely important, post-harvest management can be a difficult task for smallholder farmers in many parts of the world as compared to industrial scale agriculturists. Smallholders usually work with limited resources like manpower, financing, inputs, and others. Though challenging, small-scale agriculture is seen as more sustainable since it puts less pressure on the environment.

The instant a crop is harvested, it begins to deteriorate. Postharvest management aims at slowing this decaying process for the best quality for the consumer, either as fresh or dry produce, or as ingredients in a processed food product. The most important goals of post-harvest handling are keeping the product cool, to avoid moisture loss and slow down undesirable chemical changes (eg. grain fats could go rancid), and avoiding physical damage such as bruising or cutting, to delay spoilage. Women are often more involved than men in post-harvest operations. A gender sensitive strategy is needed to drive change in post-harvest practices and invest in new technologies.

Post-harvest losses and management depend on the type of crop. Because of its richness in proteins and sometimes lipids, grain legumes are prone to insect (pod borers in particular) and pest attacks reducing their shelf life (estimated losses of 25% in field and storage. The grain of dryland cereals, protected by husk and bran, can be stored longer.

Post-harvest management could also vary with crop genetic diversity. For instance, crop breeders are developing machine-harvestable chickpea varieties, and pearl millet cultivars with longer flour shelf life. The way grain is prepared (decortication, food preparation etc.) influences the ratio of foodgrain waste and nutrient retention. It varies according to location and even households. Sorghum in Mali is always consumed in the decorticated form (de-branning or dehulling) and the retention of minerals (eg iron or zinc) after manual decortication is as high as 44%, with variation between households.

Improving food safety is another priority. It means reducing the contact with pathogens and avoiding its multiplication because of inadequate farming and handling conditions (such as storage in non-ventilated humid conditions). Aflatoxins are fungal contamination. This naturally occurring and potentially deadly poison infests crops such as groundnuts, sorghum, cassava and maize. More than five billion people in the Majority World are exposed to aflatoxin by unknowingly consuming contaminated foods. Reducing aflatoxin contamination of African crops results in considerable health benefits.

A study shows that 37% of groundnut samples exceeded the 10 µg/kg/Kenyan regulatory limit for aflatoxin levels. The most contaminated were peanut butter and spoiled peanut (roughly three quarters) while only 4% of raw podded groundnut were contaminated. Contamination happens because appropriate storage, food processing hygiene, packaging and contamination mitigation measures are not in place. In this case it was found that 70% of storage facilities were poorly ventilated.

Post-harvest management for most crops are based on the same principles: handle with care to avoid damage (cutting, crushing, bruising), cool immediately and maintain in cool conditions, and cull (remove damaged items). Each step of the crop value chain from the field to the consumer's plate has to be carefully analyzed for quality management. Numerous biochemical processes continuously change the original composition of the crop until it becomes unmarketable. The period during which consumption is considered acceptable is defined as the 'postharvest shelf life'. The longer this is, the better the market value.

Improving value chain so that poor smallholder farmers are better linked to markets, makes them less vulnerable and better able to sell their products at higher prices. Smallholder farmers usually sell their pulse surplus immediately after harvest, while traders store pulses to take advantage of speculative gains in the off-season. Due to this, farmers do not benefit from the higher market prices of pulses.

Smallholder farmers in drylands struggle to improve the post-harvest environment which includes bad infrastructure, a challenging climate, rudimentary technologies and low capital. Healthy crops live longer and reduce post-harvest losses. A combination of capacity-building (eg training in integrated pest management to fight pod borers) and technologies (eg new legumes varieties to harvest by machine or with lower fat ranciding content, or better storage facilities with cool, dry and ventilated environment) reduces grain waste.

Aflatoxin detection kits, seed access to improved varieties, efficient finger millet grinding machines, pod-borer-free storage bags for grain legumes: solutions like these must be affordable and scalable. We need investments in market infrastructure, warehouses and market information systems that benefit smallholder farmers through Public-Private Partnership (PPP) and economically-sound models. Markets can accelerate promotion of better post-harvest management, as farmers will rapidly see the benefits of producing safer and higher quality grain.

CHALLENGES IN POSTHARVEST MANAGEMENT

Challenges for producers, traders, and handlers are always present. Here are some of the factors that can affect post-harvest management:

- Agricultural inputs and structures are hard to obtain. The lack of infrastructure, proper knowledge, and right technology can hugely affect yield and, resulting from this, income. Investments on seeds, fertilizers, and other important pre-harvest inputs can go to waste if post-harvest handling is not done properly.
- For smallholder farmers, the lack of financial support can break their livelihood. A single year's poor harvest can amount to a loss of decades' worth of savings. This can be avoided by placing secure and reliable post-harvest systems. Financial support from both non-government organizations and government bodies can also give the difference needed by these farmers to have a successful, sustainable operation.
- Gender inequalities also play a role in agricultural challenges. Women farmers are often given less assistance and opportunities than what they actually need. Land ownership, as well as access to education, financial assistance, additional labor force, and infrastructure, are all challenges to the operation and livelihood of women farmers worldwide.
- Changing technology is a big factor for post-harvest management as well. International standards for fumigation, for example, has been drastically changed due to concerns such as overuse of toxic chemicals. These shifts in technology can be difficult for producers, handlers, and traders to keep up with, especially when tied with lack of financial support.

- Another enormous challenge that the agricultural sector is facing is climate change. Adapting to the changes in weather and being prepared for natural disasters such as drought, flash floods, hurricanes, and many more can be difficult.

Effects of Postharvest Mismanagement

Mismanagement during the post-harvest season can have serious effects. Post-harvest losses occur in large-scale, with almost a third of the world's food going to loss or waste. In Sub-saharan Africa alone, more than 333 million people were affected by food insecurity in 2016, according to World Hunger Organization. At the same time, post-harvest loss greatly affects farmers' incomes. Losing a portion of their yield and the degrading quality of their commodities translate to a loss of their income. This further aggravates the impoverished status of these smallholder farmers, creating difficulties in funding their livelihood as well as in providing for their families' daily needs.

Another ill-effect of mismanagement during this stage in the agri-process is that commodities may get contaminated by debris, insects, and even mold. Such contaminants can endanger those who consume commodities, especially when substances such as aflatoxin and ochratoxin become present.

Importance of Post-harvest Management

Proper management of post-harvest systems can serve as a major help in resolving various social and economic issues. A significant decrease in post-harvest loss can alleviate food insecurity all over the world. Simultaneously, food safety can be ensured by protecting commodities from mold growth and contamination. This will result in better health and empower small communities, which can, in turn, alleviate poverty.

An increase in profits of farmers, traders, and handlers will also be seen once post-harvest loss is addressed. Through better facilities, technologies, and techniques, quality can be preserved and farmers, traders, and handlers can sell their commodities at higher prices. At the same time, loss in yield will no longer occur so more can be sold.

Having a climate-resilient and sustainable post-harvest management system can also greatly add to the success of the agricultural sector.

POSTHARVEST HANDLING

Postharvest handling activities maintain fruit quality as fresh produce travels throughout the supply chain. However, depending on the size of the farming operation and its economic situation, different specific practices are most appropriate to achieve these ends.

Post harvest technology (PHT), most popularly known as post harvest handling, is a series of methods and techniques used in the preservation of agricultural commodities after harvest. It is a science applied to agricultural commodities for the purpose of preservation, conservation, quality control and enhancement, processing, packaging, storage, distribution, marketing, and utilization to meet the food and nutritional requirements of consumers in relation to their needs.

Harvested fruits and vegetables require extensive and advance post harvest processing in order to minimize both qualitative and quantitative losses after harvesting. Every year, nearly 40% of harvested fruits and vegetables are wasted due to improper handling, storage, packaging, and transportation. As a means to counter this, post harvest technology or PHT provides trusted methods in the preservation of fruits and vegetables since they are often transported at a perishable state. These methods are used to maintain the quality of fruits and vegetables by strategically enhancing the environment related to the produce's preservation, processing, packaging, storage, and distribution. In addition to that, post harvest technology is involved not only in storage and preparation, but also extends as far as being applied to movements and operations that commodities undergo from harvest time until the very moment they are purchased by consumers.

How is Post Harvest Technology Applied?

Post harvest technology (PHT) in fruits and vegetables has three main objectives:

- To maintain quality in terms of appearance, texture, flavor, and nutritive value.
- To improve food security by safeguarding the produce's safety.
- To prevent losses involving produce and profit.

Governed by these objectives, post harvest technology is applied through minimizing rough handling, sorting to remove damaged and disease produce, and strategically taking control of the storage environment's temperature. In general, these are the main methods used to considerable minimize losses and maintain the quality of fruits and vegetables while in storage and transport.

Importance of Post Harvest Technology

Freshfruitsandvegetables(F&V)havebeenpartofhumandietsincethedawnofthehistory. The systematic nutritional value of the some F & V was recognized in the early 17th century in England. One example is the ability of the citrus fruit to cure scurvy, a diseases wide spread among naval personnel.

An example of the importance of the field to post-harvest handling is the discovery that ripening of fruit can be delayed, and thus their storage prolonged, by preventing fruit tissue respiration. The knowledge of the fundamental principles and mechanisms of respiration, leading to post-harvest storage techniques such as cold storage, gaseous storage, and waxy skin coatings. Another well known example is the finding that ripening may be brought on by treatment with ethylene. Fruits and vegetables are being rich in vitamins and minerals, known as protective foods. Due to their high nutritive valve, ready availability, and being inexpensive they make significantly contributes to human well-being. Realizing the worth of fruits and vegetables in human health ICMR recommend consumption of 120g of fruits and 280g of vegetables per capita per day.

- F&V are rich in ascorbic acid which have beneficial effects of wound healing and antioxidant. Dietary source of Vitamin C is essential, since human beings lack the ability to synthesize it.
- Some F&V are excellent source of beta -carotene (provitamin A) which is essential for the maintenance of eyes health; and folic acid which prevents anemia.
- These also prevent degenerative diseases which are prevalent in people with sedentary lifestyle. Concern about obesity and coronary heart diseases have led to reduced levels of fat intake. Antioxidants, phenolic compounds and dietary fiber are considered to be beneficial in reducing risk of various cancers.
- Many F&V have neutraceuticals properties.

F & V provide variety in the diet through difference in colour, shape, taste, aroma and texture that distinguish from the other major food groups of grains, meats and dairy products. Sensory appeal of F&V is not confined to consumption but also has market value. Diversity in their colour and shape is used by traders in arranging product displays to attract potential purchasers; and chefs have traditionally used F & V to enhance the attractiveness of the prepared dishes or table presentations; to adorn meat displays and F & V carvings have becomes an art.

The ornamental provide sensory pleasure and serenity, derived from the colors, shape and aroma of individual species. Garden plants, cut flowers, foliage and flowering plants are increasingly used in exterior and interior decoration. Considerable commercial opportunities arise from their role in social, religious and economic ceremonies and special greeting occasion such as festivals, Valentine's day and others occasion.

Total production during the year 2008-09 was of the order of 68.46 mill. ton fruits and 129.00 mill. ton vegetables and total horticultural produce was 214.71 mill. ton. Crop wise consideration shows that it has largest producer of mango, accounting to 66% of world production; holding record highest productivity in grapes; contributing to 11% of world banana and; 3rd largest producer of coconut; largest exporter of cashew nut(production + import of raw nut and then export) and 1st in spice trade.

Table: Exports of Horticultural Products (in crores).

Items	2007-08	2007-08	2007-08
Fresh Fruits	1447	1946	2269
Fresh Vegetables	1473	2454	2904
Processed Vegetables	605	711	752
Processed Fruit Juices	769	1099	1156
Miscellaneous Processed Items	1362	2077	2137

Due to inadequate post harvest handling 20-30% of horticultural produce are lost annually This loss of great magnitude not only robs labour and resources of the farmer and the nation but also dwindle away a big profit of the farmer. Managing the post harvest losses is very much important. Preservation of the produce is one of the ways to manage post harvest losses.

Table: Fruits and vegetables used for processing in different countries.

Sl.No.	Countries	%
1	Malaysia	83
2	Saudi Arabia	80

3	Philippine	78
4	Brazil	70
5	Australia	60
6	USA	40
7	UK	50
8	India	2.20

Per capita consumption is very low for our over a billion population. Major portion is being wasted at various stages of from production till it reaches end-user and its mainly due to inadequate facilities for processing. Delay in the use of harvested produce will affect its – fresh ness, palatability, appeal and nutritive value.

POSTHARVEST SYSTEM

The postharvest system spans all the operations that occur from harvest of a product to final purchase by a consumer. At one end of this system are the producers, who have grown and nurtured a product that they hope to sell at the highest possible return to their operation. At the other end are consumers, who will only purchase a product that they consider a good value.

Marketing Factors

Consumers will vary their quality perception for many reasons, for example, time of year, supply of product, supply of other products, and end use. Although initial product quality is determined by the producer, the dynamic parameters of price and quality requirements are established largely by the consumer. Any intermediary buyers (e.g., wholesalers) between the producer and the consumer also will have quality requirements.

The modern market has many established criteria; for producers to be competitive they must meet the specified requirements. Buyers will pay premium prices for fruit of uniform size and color. Fruit should not be misshapen or bruised and should be free of blemishes, diseases, and insects. For exporters, many international and national quarantine regulations must be met for insects, animals, and diseases. Product that will be stored for a length of time prior to marketing must meet criteria for maturity, firmness, and damage levels. To assist both the seller and the buyer, many public agencies and marketing organizations have developed standards for the grades of most horticultural crops. The documents commonly are called "grade standards" and include one or more sets of specifications and tolerances. Compliance with a specific set of requirements in the grade standard enables the lot to be sold as a shipment labeled with the specified grade, for example, U.S. Extra Fancy, Fancy, No. 1, or Utility. When required, a third party grade inspector evaluates samples from each lot shipped by a seller before certifying that the lot complies with the grade standard specified.

The sorting operation must be viewed in the context of the overall postharvest system. It is important to understand how cultural practices and uncontrolled inputs such as weather cause variation of quality of products that enter the packinghouse. Likewise, at the time of shipping, it is important to be able to predict the quality of shipments as they progress through the rest of the postharvest system. A major function of a packinghouse is to transform the highly variable product received from the harvesting operation into uniform lots of product for shipments that comply with the requirements of the buyer.

Terms in Sorting

Clarifying distinctions made between terms such as sorting and separation, sorter and grader, and grade inspection and quality control inspection is useful. Separation is the removal of nonusable material from usable product. An early operation in packinghouses is the separation of debris and nonedible items from the flow of marketable product. Separations are made with mechanical devices such as sizers, blowers, and washers, and by employees who remove sticks, leaves, debris, and other nonedible material.

Sorting is the segregation of edible or marketable product into distinct quality categories. Sorting of the marketable items is accomplished by both mechanical (sizers or color sorters) and human (visual or tactical) means. Sorters also may perform some separation tasks such as removing debris and nonmarketable materials missed by earlier separation operations. The equipment used for sorting should be referred to as a sorting line, not a grading line. People who perform sorting operations should be referred to as sorters, not graders. Graders are the third party inspectors who evaluate whether or not the packed lot complies with requirements of a grade standard for a predetermined grade classification.

Sample inspection for quality control is more precise than, and differs from, the inspection of product necessary for sorting. The importance of the sorting operation cannot be overstated, since variations in this operation will affect returns for most other parts of the postharvest handling system.

Sorting Equipment

Many attempts have been made to automate the sorting operation using machine vision. The primary impediments to progress are the complexities of defining algorithms to describe the numerous defects that must be identified by an image processor. One review indicates that progress has been made for selected products and features. The economics of machine vision continue to hinder wider applications, especially for packinghouses with short operating seasons. Some systems have been developed that combine human skills of vision and defect detection with mechanical removal.

Most sorting operations still are performed by human visual inspection of the product and manual removal of items with defects. Humans have special abilities for identifying defects and for determining if they exceed prescribed threshold criteria. This combination of rapid detection and flexibility in selecting thresholds for rejection suggest that postharvest systems will always include some form of human visual sorting operation. However, special attention is needed to avoid inconsistent performance.

A typical sorting operation consists of a continuous flow of product passing in front of one or more stationary sorters. Normally, the task of the sorter is to remove items that do not meet the specifications for the lot being shipped. Nonconforming items are placed into a discard flow and items meeting other specifications are placed on separate conveyers that may flow to packing areas for lesser quality markets.

The design of sorting equipment has considerable effect on efficiency of the sorter in detecting and removing defective items that cause off grades. Consideration must be given to how the product is presented to the sorters. For efficient operation, the product must be easy to see, reach, and remove, and reject chutes or belts must be readily accessible.

For horticultural crops, a sorting conveyor is the customary means for presenting products to inspectors; four common techniques are shown in figure below. When possible, items are rotated, which makes it easier for sorters to see defects over the entire product surface. A common sorting table for items with a spherical or cylindrical shape consists of a roller conveyor supporting the items between consecutive rollers. In many designs, the rollers are dragged over a flat surface. This friction surface rotates the rollers and, consequently, the product. The drive surface may be stationary, or driven to enable variable control of product rotation speed.

Alternatively, the "push-bar conveyor" pushes fruit over a flat surface. The push-bars are a set of stainless steel or plastic tubes separated by at least the diameter of the product. Drive chains attached to both ends of the bars pull them along the table surface, causing them to rotate the product as it is pushed past the sorters. The sorting surface may be fixed, but a flat belt is used often to enable the rotation speed of the product to be varied.

The shape or size of some products precludes the option of easily rotating them; in these cases a flat belt conveyor is used. This technique is common for small fruit such as cherries and berries. The major disadvantage of this type of presentation is that the product can be viewed only from above; any defects on the underside are not visible unless the fruit is moved manually.

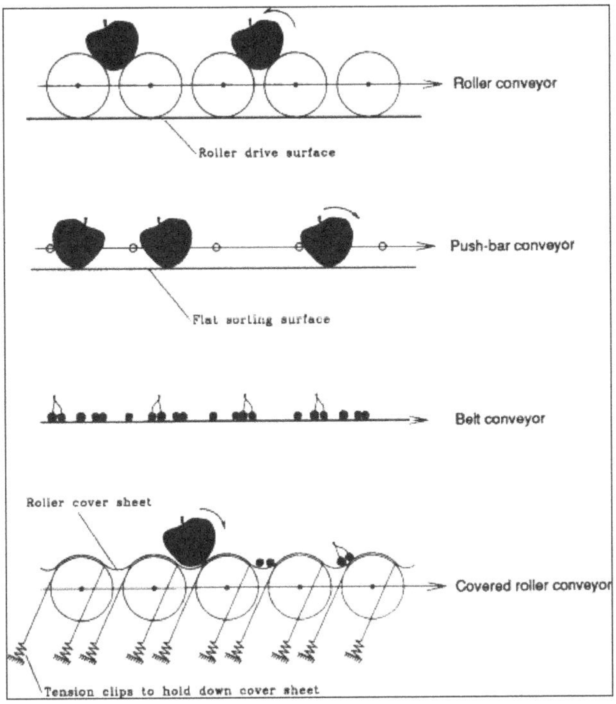

Typical methods for transporting product past sorters.

Fruits can be rotated by cascading them from one belt to a second, where the product is reinspected, or by placing curtains over the table which drag over the product and rotate it. The height of the curtains can be adjusted to enable most of the product to be rotated by 180°. A combination design that uses a roller conveyor with a flat surface is shown in figure above also. This approach uses a fixed roller cover sheet that is held tightly over a roller conveyor with a series of tension clips. The movement of the rollers under the sheet causes a ripple effect that propagates along the table. The product is transported in the valleys, and rotates as it moves forward.

Visual Perception

The ability of humans to perceive a visual image depends on both physical and cognitive factors. Changes in the color and intensity of light change the image received by the eye. The method of presenting the product to the sorters also has an important effect on perception. If product speed (either translation or rotation) is too fast, it is not possible to fixate properly on a defect; hence, it is not possible to reach a decision about whether or not the item should be rejected.

Any vision difficulties adversely influence detection of defects. Visual acuity decreases with age, but can be increased by increasing the brightness of the test object. A 60-year-old worker requires about twice the brightness level that a 20-year-old worker requires for equal visual acuity. Vision examinations for sorters are useful for determining problems with visual acuity, peripheral vision, and color blindness. Also, the inability to concentrate for long periods of time results in a relaxing of vigilance, which is an important factor of visual perception.

The postharvest system can be split into three parts, in terms of traceability. At the packinghouse, product is placed into packaging and onto pallets. From this stage in the process product is well identified and can generally be traced through the logistics chain. Through this part of the system it is possible to achieve complete traceability, provided adequate recording of activities and product movement is carried out. Once the pallets and packs are broken down, traceability becomes less accurate as product is mixed and the systems used tend to be less robust. This is also the case from the orchard until the point at which the product is packed. So, the system is characterized by a central component of high traceability sandwiched between two components of low traceability. While it is common in discussions of traceability to assume absolute traceability, in fact in the context of the entire postharvest system, traceability will never be absolute.

A traceability system can be described by different sizes of IU, which is termed the granularity of the traceability. At the coarsest level of granularity, a fruit or vegetable may be traceable to a state or country of origin, or perhaps an individual farm or orchard, whereas finer levels include individual shipments, pallets or packs. The ultimate level of traceability would be individual fruit, but currently there are no practical systems that enable traceability at this level.

The processing operations such as cleaning, sorting, grading of fruit in a packing house involve taking fruit from definable IUs, such as individual bins picked in the orchard, and handling these as a continuous stream. The packing house generally, but not always, splits the fruit into batches of common product (from the same orchard, or orchard block). These batches are packed in sequence through the facility and tracked onto the final pallets. Often there may be different grower's fruit on the same pallet and the individual growers are generally identified on the packs, but in many cases this is simply a generic grower ID that is not specifically linked to a batch. There are also likely to be packs at the changeover from one batch to another that are filled with fruit from both batches. Traceability in this case relies on both the pallet IU, which will generally be linked to a batch, as well as having grower IDs associated with this IU. Trace back, however, would require manual identification of individual packs as there is no potential to link these into a traceability information system without a unique pack ID.

PLANT AND POSTHARVEST PHYSIOLOGY

Postharvest physiology is about the plant response to technologies and other applications that extend shelf life and quality and delay senescence (plant death). Quality itself includes aspects that are appealing to consumers, such as appearance, aroma, taste, color (sensory), reduced risk of foodborne pathogens or pesticide residues (food safety) or more dense nutrient or phytonutrient content. Growers and shippers realized that keeping produce or flowers cold greatly extended the sales life and this stimulated much interest in exploring the causes of plant senescence at physiological, cellular and molecular levels. As consumers continue to increase their daily intake of fruits and vegetables, pressure to deliver high quality product with a longer shelf life is critical.

Despite the remarkable progress made in increasing food production worldwide, approximately half of the population in the developing countries does not have access to adequate food supplies; thus the food security problem is worsening. There are many reasons for this, one of which is food losses occurring throughout the supply chain from production, post-harvest, processing and marketing. Roughly one-third of food production for human consumption is lost or wasted globally, which amounts of about 1.3 billion tons per year. Evidence suggests that these losses tend to be highest in those countries where the need for food is greatest.

Pre-harvest conditions and events in the fields such as cultivar, soil type, fertilization and irrigation practices, weather conditions, pest control programs etc. have a significant influence on achievement of the best potential postharvest quality and shelf life of fresh fruits and vegetables.

Plants or plant parts continue to function metabolically after harvest and are subjected to physiological and pathological deterioration and loss. “Loss” means any change in the quality of the food that prevents it from being consumed by people. Postharvest loss in fresh fruits and vegetables is estimated at 5-25% in developed countries and 20-50% in developing countries. Causes of this loss are varied with microbiological, mechanical and physiological factors being the main cause in perishable crops. Other causes are in-adequate harvesting, packaging, handling skills and refrigerated storage, as well as inadequate transportation.

Storage and shelf life are defined as the period from harvest to consumption, while a food product remains safe and wholesome. There is a wide range of postharvest technologies that can be adopted to reduce losses throughout the process from field to fork. Both quantitative and qualitative food losses of extremely variable magnitude occur at all stages in the post-harvest system from harvesting, through handling, storage, processing and marketing to final delivery to the consumer. Appropriate storage can minimize moisture loss, slow down respiration rate and inhibit development of decay-causing pathogens. Wilting, re-growth, ripening, senescence and decay can be postponed. Temperature is the most important determination of fresh produce deterioration rate. An important supplement to temperature and relative humidity management is the use of controlled atmosphere (CA) or modified atmosphere (MA) and other technologies.

Postharvest loss results not only in the loss of the actual crop, but also have an impact on the environment, resources, labor needed to produce the crop and livelihood of individuals involved in the production process. The implementation of appropriate storage and postharvest techniques will add value to the produce and will increase the farmer’s income.

References

- The-importance-of-post-harvest-technology-for-fruits-and-vegetables, crops: bionovelus.com, Retrieved 21 March, 2020
- Postharvest-management: exploreit.icrisat.org, Retrieved 30 May, 2020
- Postharvest-systems, agricultural-and-biological-sciences: sciencedirect.com, Retrieved 21 July, 2020
- Plant-postharvest-physiology, areas, research, horticultural-science: cals.ncsu.edu, Retrieved 3 January, 2020
- Postharvest-physiology, mashav, yaounde: embassies.gov.il, Retrieved 22 June, 2020

2

Postharvest Processing

Postharvest processing is a crucial stage of postharvest management. It helps in preventing spoilage of crops. Some of the processes of this domain are ultraviolet light treatment, pre-cooling, fruit waxing, drying, sorting and grading, etc. This chapter explores these processes to provide an extensive understanding of the subject.

Postharvest processing is a continuum of three crucial primary, secondary, and tertiary steps in adding value and enhancing utilization of dry beans. Postharvest processing (drying and cleaning) is very important for maintaining high-quality grain. Problems with grain drying will vary from year to year depending on weather conditions. Grain moisture of 10–12% is optimum for storage without further drying. It is important to harvest the grain when the plants are as dry as possible. The longer the plants stay in the field after the first frost, the higher the likelihood that the quality of the grain will be reduced by wet weather.

To prevent recontamination during postharvest processing, produce should not come into contact with any reservoir of pathogenic E. colie specially manure-amended soil and water. Removal of residual soil during the postharvest processes can be an effective way of reducing the pathogen load on the surface. However, water can contaminate both the surface and the inner part of the produce. Some human pathogens, including E. coli O157:H7, can penetrate or infiltrate the interior of produce and are thus more resistant to sanitizing agents. This internalization phenomenon occurs more easily when produce is damaged, either naturally or artificially (e.g., by trimming, cutting, and chopping).

Various studies have examined penetration and internalization of pathogenic E. coli using different technologies (e.g., direct tissue printing, immunoblot tissue printing, SEM, TEM, confocal microscopy, immune localization, and light microscopy); however, the mechanism remains unclear. Among the different sources of contamination, washing with dirty water presents pathogenic E. coli with an internalization route into lettuce, oranges, and tomatoes. Unhygienic washing water is a major source of contamination and should be avoided at all times. For this reason, adequate disinfection (e.g., by electrochemical treatment and ultrasonic irradiation) of recycled water is recommended.

POSTHARVEST TREATMENTS OF FRESH PRODUCE

Fresh fruit and vegetables (FFV) are a major source of essential vitamins and minerals, such as vitamin A, vitamin C and potassium, needed for human wellbeing. They are, however, perishable living products that require coordinated activity by growers, storage operators, processors and retailers to maintain quality and reduce food loss and waste. The extent of coordination can vary greatly from loose in the case of local food supplies too complex for global supply chains. The Food and Agriculture Organization estimated that 32% (weight basis) of all food produced in the world was lost or wasted in 2009. When converted into calories, global losses represent approximately 24% of all food produced. Reducing the loss and waste of FFVs is important because these foods provide essential nutrients and represent sources of domestic and international revenue.

Fresh produce attributes (appearance, texture, flavour and nutritional value) have been traditional quality criteria, but increasingly safety (chemical, toxicological and microbial) and traceability are important for all the role players along the supply chain, from the farm to consumers. Fresh produce is often eaten raw or after minimal processing and food pathogen contamination can present risk of outbreaks of foodborne illnesses. Listeria monocytogenes, Salmonella enteritidis phage, and Escherichia coli O157:H7 and O104:H4 are major pathogens contributing to outbreaks of foodborne illness with fresh produce as vectors for these pathogens. Owing to multiple uncertainties along the supply chain, microbial contamination leading to spoilage and postharvest losses can occur at any of the stages in the continuum from farm to consumer. Therefore, postharvest treatments are essential to minimize microbial spoilage and reduce the risk of pathogen contamination for FFV.

Various postharvest physical, chemical and gaseous treatments may be applied to maintain fresh-like quality with high nutritional value and meet safety standards of fresh produce. These postharvest treatments are typically combined with appropriate management of storage temperatures.

Physical Treatments

Heat treatment has been studied as an alternative to chemical treatments for harvested FFV. Treatments include hot water dip (HWD), saturated water vapour heat, hot dry air and hot water rinse (HWR) with brushing. Beneficial effects of these heat treatments are linked (i) through changes in physiological processes such as a reduction in chilling injury and delay of ripening processes by heat inactivation of degradative enzymes, (ii) by killing of critical insect contaminations, and (iii) by controlling the onset of fungal decay. Heat treatments can be of short- (up to 1 h) or long-term duration (up to 4 days). Heat treatments have been applied to firm potatoes, tomatoes, carrots and strawberries; to preserve the colour of asparagus, broccoli, green beans, kiwi fruits, celery and lettuce; to prevent development of overripe flavours in cantaloupe and other melons; and to generally add to the longevity of grapes, plums, bean sprouts and peaches, among others.

It has been demonstrated that heat shock by using hot water washing at temperatures ranging from 37 to 55 °C for a duration of 30 s to 3 min can improve the postharvest quality of spinach, rocket leaves, apples and mandarin fruit. A clear mode of action of any water treatment is to wash-off the spores from the fruit surface. Hot water is a better vector of energy than air and has provided comparable reductions in fungal decay. Blue mould on grapefruit caused by Penicillium sp. has been controlled by dipping fruit in hot water for 2 min at 50 °C. Improvements in the quality of bell pepper, apples, melons, sweet corn, kumquat and grapefruit have been reported with cold water cleaning in combination with brushing and a short HWR. Hot water treatments also influence the structure and composition of epicuticular waxes. Covering of cracks and wounds and the formation of anti-fungal substances in the wax after heating are thought to be possible modes of action.

The combination of Bacillus amyloliquefaciens HF-01, sodium bicarbonate and hot water could be a promising method for the control of postharvest decay on citrus while maintaining fruit quality after harvest. To date, commercial applications of heat treatments are limited. Heat treatment provides an alternative to fungicide applications, and in Germany HWD has been used in the storage of organic apples. Treatment of fruit after a few days of cold storage or immediately after the opening of a long-term controlled atmosphere (CA) storage room provides new options for prolonging their subsequent storage life, although acceptance of this technology by fruit growers has been hampered by high-energy costs and also the need for added labour at the peak work period during harvest time.

Edible Coating

Edible coatings are thin layers of external coatings applied to the surface of fresh produce to enhance the waxy cuticle or as replacements for natural barriers where the produce cuticle has been removed. The application of edible coatings on fresh produce provides a partial barrier to the movement of moisture on the surface of fresh produce, thereby minimizing moisture loss during postharvest storage; a gas barrier, thereby establishing a modified atmosphere around the product, which slows down respiration, senescence and enzymatic oxidation and preserves colour and texture; helps to retain volatile compounds contributing to produce a natural aroma and restrict foreign odours; maintains fresh produce structural integrity, and protects against mechanical damages; and serves as carriers of functional or active compounds, such as nutraceuticals, flavouring and colouring agents, antioxidants and antimicrobials, that will maintain/improve product quality and safety. Edible coatings are composed of hydrophobic groups, such as lipid-based waxes; hydrocolloid/hydrophilic groups, such as polysaccharide or protein-based materials; or an integration of both groups in order to improve the functionality of the coating. Within the last decade, there has been a considerable amount of research and innovations focused on the development of edible coatings from natural or synthetic sources in order to control physiological and pathological challenges of fresh produce.

Table: Edible coatings used on fresh/fresh-cut fruit and vegetables.

Coating material	Purpose of coating
Guar gum; pea/potato starch ± potassium sorbate	Antimicrobial
Candelilla wax-based	Antimicrobial; antioxidant; quality
Soya bean gum; jojoba wax; glycerol and arabic gum	Overall quality
Shellac ± Aloe vera gel	Keeping quality
Soy protein; carboxymethyl cellulose	Antioxidant; H_2O barrier
Chitosan; zein	Antioxidant; H_2O barrier
Beeswax; coconut and sunflower oil	Antimicrobial; antioxidant; quality
Pectin base; alginate; carboxymethyl cellulose	Antioxidant; H_2O barrier
Chitosan; methyl cellulose	Antimicrobial; antioxidant; $O_2/CO_2/H_2O$ barrier
Soy protein; carboxymethyl cellulose	Antioxidant; H_2O barrier
Pectin base	Overall quality
Aloe vera gel	Overall quality
Agar; chitosan; acetic acid (combined)	Antimicrobial; O_2/CO_2 barrier
Whey protein; rice bran oil	H_2O barrier; overall quality
Chitosan	Overall quality
Sucrose-polyester based	H_2O barrier; antioxidant activity
Alginate and gellan based	$O_2/CO_2/H_2O$ barrier

Several edible coatings including chitosan, Aloe vera, polyvinyl acetate, mineral oils, cellulose and protein based have shown desirable attributes on fresh produce with good barrier properties, without residual odour or taste and efficient antimicrobial activity. However, more research is required to enhance moisture barrier properties of hydrophilic edible coatings, improve coating adhesion and durability during storage. To maximize the benefits of edible coatings for fresh produce, it is important to understand the effect of storage conditions on the desired functions and the adverse effect on fresh produce quality. The main limitation for the application of edible coating at the industrial level is the cost of scaling up research concepts or investment for new installation of film production and coating equipment, the lack of edible materials with desired physical and functional properties as well as the challenges of regulatory status for the different coating materials. Furthermore, process parameters, such as the method of coating and the amount of additives, can affect the film barrier properties and overall quality of the food product. One of the commercial coating products is Natureseal, which maintains colour, texture and shelf life of a number of fresh-cut fruits e.g. apples, pears, carrots, celery, etc., has recorded a good success. However, further research development is required to investigate the influence of edible coatings on individual cultivars of fresh-cuts in order to understand the variation in shelf life.

Irradiation

Irradiation exposes food to radiant energy from γ rays and e-beam (high-energy electrons) that penetrate objects and break molecular bonds, including the DNA of living organisms. Ionizing radiation from cobalt-60 or caesium-137, or machine generated electron beams are used as a source of irradiation for extending shelf life of fresh produce. By inhibiting cellular reproduction, irradiation can neutralize pest and food safety problems. The effect depends on the doses, measured in kilograys (kGy). Low doses of irradiation (less than 1 kGy) only disrupt cellular activity enough to inhibit sprouting of tubers, bulbs and roots and delay senescence. Medium doses (1–10 kGy) reduce microbial loads while high doses (more than 10 kGy) kill a broad spectrum of fungi and bacteria spp. and pests. Most medium- and high-level doses are not appropriate for fresh produce because they can cause sensory defects (visual, texture and flavour) and/or accelerated senescence due to the irreparable damage to DNA and proteins. Irradiation presents an effective postharvest treatment for destroying bacteria, moulds and yeasts, which cause food spoilage, and also control insect and parasite infestation resulting in reduced storage losses, extended shelf life and improved parasitological and microbiological safety of foods. Irradiation has been commercialized for control of potato and onion sprouting, and strawberry decay. Low-dose γ-irradiation on mango (0.3–0.7 kGy) resulted in delay in ripening and extension of shelf life by a minimum of 3–4 days. Recently, Pandey et al. reported an irradiation dose of 1 kGy to be the only effective dose in which enhanced shelf life was achieved without any deterioration of various quality attributes of litchi fruit.

While much of the focus of irradiation use on FFV has been for extending shelf life and reducing decay, it has been known for many decades that irradiation is effective at killing, sterilizing or preventing further development of a wide variety of insect pests of quarantine importance on perishable FFV. Despite some misconceptions, exposing food to irradiation does not make the food itself radioactive. The irradiation process produces very little chemical change in food and does not change the nutritional value of food. Extensive research and testing has demonstrated that irradiated food is safe and wholesome.

Chemical Treatments

Antimicrobial and Anti-browning Agents

Over the past decade, the increasing number of reported outbreaks of foodborne illnesses has heightened the concern of regulatory agencies, producers and the consumers about the microbial safety of FFV. Outbreaks have been associated with vegetables such as cabbage, celery, cucumber, leeks, watercress, lettuce and sprouts.

Antimicrobial and anti-browning agents offer the possibility to maintain safety and can be grouped into chemical- and natural/bio-based agents. Chemical-based agents include chlorine-based solutions, peroxyacetic acid (PAA), organic acids, hydrogen peroxide (H_2O_2) and electrolysed water. A chlorine-based solution such as NaClO has been one of the commonly used disinfectants for fresh produce, owing to its very potent oxidizing properties and cost effectiveness. However, its efficacy as an antimicrobial agent is dependent on the levels of chlorine and at high levels may cause taste and odour defects on treated products. Additionally, chlorine-based compounds have been reported to have limited effectiveness in the reductions of microbial load on fresh produce. Surfactants, detergents and solvents, alone or coupled with physical manipulation such as brushing, may be used to reduce hydrophobic nature of the waxy cuticle or remove part of the wax to increase exposure of microorganisms to chlorine. However, chlorine has been associated with the possible formation of carcinogenic chlorinated compounds and this may lead to new regulatory restrictions in the EU.

PAA is a very strong oxidizing agent, with no harmful by-products. PAA has been reported to be effective in controlling E. coli O157:H7 and L. monocytogenes on apples, strawberries, lettuce and cantaloupe. A 5 log reduction in Enterobacter sakazakii was reported for lettuce when treated with PAA. Landfeld et al. reported that decontamination treatment of fresh-cut carrot with PAA reduced the initial load of aerobic mesophilic bacteria by about 4 log units and yeasts and moulds by 3.5 log units and no further microbial growth was observed during storage.

H_2O_2 possesses a bactericidal, sporicidal and inhibitory ability, owing to its property as an oxidant and being able to generate other cytotoxic oxidizing species, such as hydroxyl radicals. Treatment with H_2O_2 can extend the shelf life and reduce natural and pathogenic microbial populations in melons, oranges, apples, prunes, tomatoes, whole grapes and fresh-cut produce. However, H_2O_2 treatment requires a long duration of application and can cause injury on some produce. Also, it is accepted as a generally recognized as safe for some food applications but not yet approved as an antimicrobial agent. However, a recent study found that the newly developed H_2O_2-based sanitizers provoked a significant increase in the respiration rate and the electrolyte leakage of fresh-cut iceberg lettuce compared with tap water washing.

Organic acid, ascorbic acid and calcium-based solutions have been applied largely to slow down enzymatic and non-enzymatic browning, deterioration of texture and microbial growth on fresh produce. Treatment of fresh-cut melon dipped in 0.52 mM citric acid for 30 s prior to modified atmosphere packaging (MAP) maintained microbial safety and prevented translucency and discoloration. Inhibitory effects of organic acids (acetic, lactic and malic acids) combined with MAP on foodborne pathogens, including E. coli O157:H7, S. Typhimurium and L. monocytogenes, on cabbage was reported by Bae et al. However, there are factors limiting the efficacy of antimicrobial and anti-browning agents such as internalization of bacteria and inaccessible sites within fresh produce such as the calyx. These limitations highlight the need for novel means of applying of antimicrobial and anti-browning agents.

Nitric Oxide

Nitric oxide (NO) is a highly reactive free radical gas and acts as a multifunctional signalling molecule in various plant physiological processes, such as fruit ripening and senescence of FFV. Endogenous NO concentrations decrease with maturation and senescence in FFV, thereby offering an opportunity for modulation of their levels with exogenous application to exert the opposite effects. Optimum NO levels delay the climacteric phase of many tropical fruits, prolong the postharvest storage life by impeding ripening and senescence, suppress biosynthesis of ethylene, reduce ethylene production and, consequently, delay in fruit ripening. NO gas is applied as a fumigant or released from compounds such as sodium nitroprusside, S-nitrosothiols and also diazeniumdiolates used as a dipping treatment. Reduced ethylene production during ripening in NO-fumigated fruit has been claimed owing to binding of NO with 1-aminocyclopropane-1-carboxylic acid (ACC) and ACC oxidase to form a stable ternary complex, thus limiting ethylene production. Other mechanism of NO action include the inhibition of ethylene biosynthesis, cross-communication with other phytohormones, regulation of gene expression and amelioration of oxidative postharvest stress.

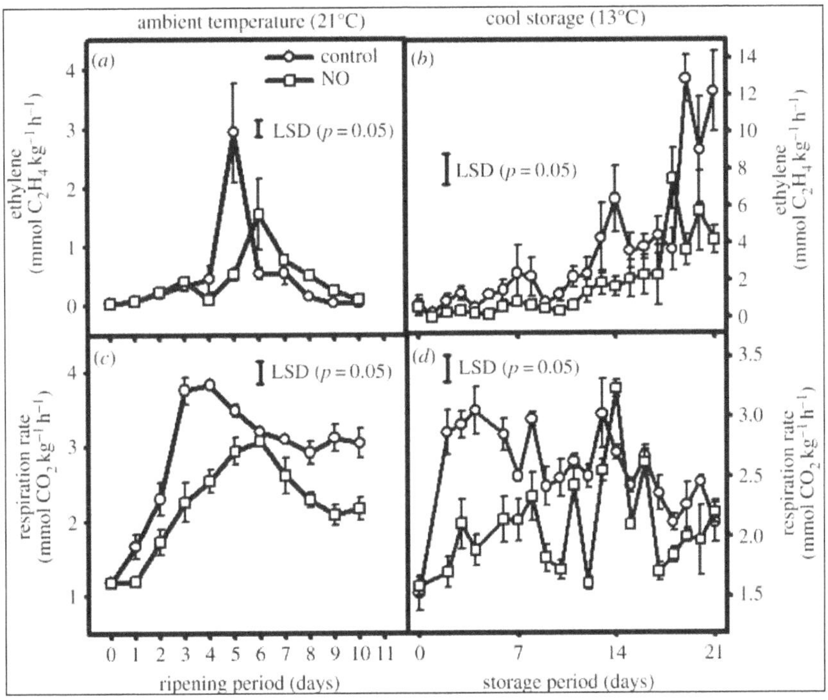

Ethylene production (a,b) and respiration rate (c,d) of mango as influenced by NO fumigation, ripening period and storage temperature.

Successful application of NO has been reported for apple, banana, kiwifruit, mango, peach, pear, plum, strawberry, tomato, papaya, loquat, Chinese winter jujube fruit and

Chinese bayberry. NO treatments reduce the rate of respiration, water loss and inhibit browning, as well as reducing the incidence of postharvest diseases. NO fumigation in combination with cold storage has a synergistic effect in extending storage life of fruit such as plum and mango. Similarly, NO combined with modified atmosphere conditions extended the postharvest life of green beans, broccoli and bok choy. Commercial application of NO in FFV depends upon the development of a smart carrier/controlled release system for NO.

Sulfur Dioxide

Sulfur dioxide (SO_2) is widely used on table grapes to prevent decay during storage, by either initial fumigation of fruit from the field followed by weekly fumigation of storage rooms or slow release from in-package pads containing sodium metabisulfite. SO_2 technology has also been tested for control of postharvest decay on other fruits such as litchi, fig, banana, lemon or apple. Cantín et al. reported that SO_2 fumigation followed by CA storage (3%O_2+6 or 12%CO_2) is a promising postharvest strategy for fresh blueberries to reduce decay, extend market life and maintain high nutritional value. Rivera et al. demonstrated that SO_2 is an effective and practical technology for reducing the risk of blueberry grey mould decay during storage and it could be used for the export market. There are disadvantages to SO_2 use; the SO_2 concentration necessary to inhibit fungal growth may induce injuries in grape fruits and stems, and sulfite residues pose a health risk for some individuals, as well as firming of the texture of some fruit species (pomaces), incomplete de-sulphiting and incomplete re-colouring of red fruits. Nevertheless, SO_2 treatment is a widespread process because of its advantages of universal antiseptic action and economic application.

Gaseous Treatments

Ozone

Recent research and commercial applications have verified that ozone can replace traditional sanitizing agents. Ozone is a very pungent, naturally occurring gas with strong highly reactive oxidizing properties. Ozone is reported to have 1.5 times the oxidizing potential of chlorine and 3000 times the potential of hypochlorous acid. Contact times for antimicrobial action are typically four to five times less than that for chlorine. Ozone rapidly attacks bacterial cell walls and is more effective than chlorine against the thick-walled spores of plant pathogens and animal parasites, at practical and safe concentrations. Ali reported that the fruit exposed to 2.5 ppm ozone had higher levels of total soluble solids, ascorbic acid content, β-carotene content, lycopene content, and antioxidant activity and also reduced weight loss at day 10 compared with untreated fruit. The sensory attributes of papaya of ozone-treated fruit was also superior in sweetness and overall acceptability endorsing ozone as a non-thermal and safe food preservation technique for FFV. Ozone can be employed in cold storage, washing system or process water sterilization. Huyskens-Keil et al. reported that irradiation and washing with ozonated water slightly reduced respiration in white asparagus spears, but increased spear tissue toughness. However, neither washing the asparagus spears with ozonated water (3 or 4.5 ppm) nor treating them with radiation (1 kJ m^{-2}) systematically and significantly affected their microbial loads during storage. Some commercial use has occurred with commodities such as apples, cherries, carrots, garlic, kiwi, onions, peaches, plums, potatoes and table grapes. However, ozone does not penetrate natural openings or wounds efficiently. Additional research is needed to define the potential and limits of the effective use of ozone for postharvest treatments for the quality and safety of FFV.

Ethylene

Endogenous ethylene production and its exogenous application exhibit both beneficial and deleterious effects on horticultural fresh produce. Beneficial effects of exogenously applied ethylene includes triggering ripening, improving fruit colour and quality in some crops, such as bananas and avocados, kiwifruit, persimmon, tomato, mangoes, de-greening of citrus fruit. The deleterious effects of ethylene in postharvest phase horticultural commodities has also been documented, such as shorter storage life, promotion of senescence, fruit softening, discoloration (browning) and russet spotting in lettuce, yellowing of leafy vegetables and cucumbers and increased susceptibility of FFV to decay. Therefore, ethylene management plays a pivotal role in maintaining postharvest life and quality of climacteric and non-climacteric horticultural produce. Most commercial strategies for maintaining horticultural commodities involve storing at low temperatures, blocking ethylene biosynthesis and its action, minimizing exposure produce to ethylene during ripening, harvest, storage and transport by controlling temperature and atmospheric gas composition. Newly developed ethylene measurement devices will enable to detect critical concentrations during storage and transportation.

Beneficial effects of ethylene biosynthesis inhibitors such as aminoethoxyvinylglycine alone on postharvest quality have been demonstrated in apples and stone fruits and in combination with controlled atmosphere (CA) storage. Treating FFV with inhibitors of ethylene action, such as 1-methylcyclopropene (1-MCP) or NO alone or in combination with MAP or CA storage, also impedes ethylene production and action consequently extends storage life and maintains quality of FFV.

A recent development has been palladium-promoted zeolite materials, which can be effective ethylene scavengers to prolong the shelf life of climacteric fresh produce, such as bananas and avocados. The material has the potential to be used commercially, as an alternative and/or supplemental treatment to 1-MCP. Martínez-Romero et al. developed a carbon-heat hybrid ethylene scrubber for fresh horticultural produce storage purposes. The device comprised a cartridge heater tightly joined to the activated carbon–1% palladium. Application of heat pulses leads to an increase in ethylene oxidation and to auto-regeneration of the activated carbon.

1-Methylcyclopropene

The discovery and patenting of cyclopropenes as inhibitors of ethylene perception represents a major breakthrough in controlling ethylene responses of horticultural products. Of the cyclopropenes, 1-MCP proved to be extremely active, but unstable in the liquid phase. However, 1-MCP can be complexed with α-cyclodextrin to maintain its stability; this development represented a major step towards its commercialization as it was then possible to release 1-MCP from the complex to expose to the horticultural products. Regulatory approval for use of 1-MCP has been obtained in more than 50 countries, and approval for use of the technology continues to occur around the world. 1-MCP is registered for use on a wide variety of FFV including apple, avocado, banana, broccoli, cucumber, date, kiwifruit, mango, melon, nectarine, papaya, peach, pear, pepper, persimmon, pineapple, plantain, plum, squash and tomato. 1-MCP affects many ripening and senescence processes, including pigment changes, softening and cell wall metabolism, flavour and aroma, and nutritional properties, but to varying degrees in both non-climacteric and climacteric products. While aqueous 1-MCP shows similar responses as those treated with gaseous 1-MCP, ripening actors such as activity of cell wall-associated enzymes, e.g. lycopene, antioxidant and volatiles of avocado, are delayed but recover to reach levels similar to those of untreated fruit. The range of responses reflects the enormous diversity of these crops in terms of both inherent diversity and morphological derivation. Several generalizations can be made about responses of crops to 1-MCP:

- Genotype, cultivar and maturity effects can be highly variable.
- Climacteric fruit are affected by 1-MCP treatment, but the capacity to interrupt ripening once initiated varies by fruit and attributes studied.
- Non-climacteric fruit also respond to 1-MCP, providing insights into ethylene-dependent and -independent events during ripening.
- Treated fruits are firmer, slower to soften, slower to change peel colour, and develop aroma and flavour. If 1-MCP concentrations and exposure periods for each fruit must be appropriate to allow ripening to occur, so the final quality of the treated fruit is similar to that of the untreated product.
- The effect of 1-MCP on physiological disorders is dependent on the role of ethylene. 1-MCP decreases senescent-related and ethylene-induced disorders (senescent breakdown of apples and water soaking of watermelons). 1-MCP can increase susceptibility of apple fruit to CA storage-related disorders (CO_2 injury). Some chilling-related disorders (woolliness and internal breakdown of peaches and nectarines, chilling injury of citrus and bananas) can be increased by inhibition of ethylene production. Other chilling-related disorders (superficial scald of apples and pears and internal flesh browning of avocados and pineapples) are decreased by inhibition of ethylene production.

Registration of 1-MCP for FFV has focused on major or specialty products important to specific countries. The apple has been an excellent crop for use of 1-MCP, and the technology is used extensively around the world to maintain quality through the whole marketing chain from storage to consumer. Success of 1-MCP technology for apples is largely associated with a fruit where maintenance of 'at harvest' quality and only moderate softening to a crisp fracturable texture is desirable. Watkins et al. found that rapid treatment of fruit with 1-MCP after harvest can afford storage operators more freedom to delay CA storage application, but attention to cultivar, fruit maturity and susceptibility of fruit to storage disorders must be considered. Challenges exist for the effective use of 1-MCP for fruit that ripen uniformly to a melting texture and/or have major colour, flavour and aroma changes that are expected by the consumer. Failure to ripen normally has been shown in avocado, banana, pear and tomato, where fruits were treated at an early ripening stage or where applied 1-MCP concentrations were too high. Despite the challenges, successful commercialization of SmartFresh for treatment of avocados, bananas, melons, persimmons and tomatoes has resulted from careful attenuation of 1-MCP concentrations and/or ripening stage at harvest.

Recently, 1-MCP formulations have been approved by the Environmental Protection Agency and other regulatory authorities for preharvest applications, and these are marketed as Harvista for FFV. Semi-commercial trials have been carried out in several US locations, Argentina, Brazil, Canada, Chile, New Zealand and South Africa. Harvista has useful effects on delaying fruit drop, slowing fruit maturation and ripening, and in maintaining postharvest quality, including improving the effects of SmartFresh.

Controlled Atmosphere Storage

CA storage refers to the monitoring and adjustment of the CO_2 and O_2 levels within gas tight stores at optimum storage temperature. Thus, the atmosphere is controlled rather than established passively as in the case of MAP, though the effects of altered atmospheres on metabolism of FFV are essentially the same. In most cases, the concentrations of CO_2 are higher and those of O_2 are lower, optimum concentrations depending on the specific product and the purpose of the CA storage conditions. Reduced O_2 and elevated CO_2 levels affect both primary (glycolysis, fermentation and aerobic respiration) and secondary (e.g. processes involved in ethylene production and action, pigments, phenolics and volatiles) metabolism.

Each FFV has an optimal range of O_2 and CO_2 for maintaining quality and extending shelf life, and these can differ for whole and fresh-cut products of the same fruit or vegetable. Benefits of CA as well as detrimental effects resulting from exposure of FFV to atmospheres outside of the safe range were summarized by Kader. The beneficial effects of CA include:

- Retardation of senescence and associated biochemical and physiological changes, e.g. slowing down rates of respiration, ethylene production and softening.

- Reduction in sensitivity to ethylene action at O_2 levels less than 8% and/or CO_2 levels more than 1%.
- Alleviation of certain physiological disorders, e.g. chilling injury of avocado and superficial scald of apples and pears.
- Direct or indirect effect on postharvest pathogens (bacteria and fungi) and consequently decay incidence and severity e.g. CO_2 at 10–15% inhibits development of Botrytis rot on strawberries and cherries.
- Low O_2 (less than 1%) and/or elevated CO_2 (40–60%) can be a useful tool for insect control in some fresh and dried fruits, flowers and vegetables; and dried nuts and grains.

The detrimental effects of CA include:

- Initiation and/or aggravation of certain physiological disorders, such as internal browning in apples and pears, brown stain of lettuce and chilling injury of some commodities.
- Irregular ripening of fruits, such as banana, mango, pear and tomato, can result from exposure to O_2 levels below 2% and/or CO_2 levels above 5% for more than one month.
- Development of off-flavours and off-odours at very low O_2 concentrations (as a result of anaerobic respiration) and very high CO_2 levels (as a result of fermentative metabolism).
- Increased susceptibility to decay when the fruit is physiologically injured by insufficient O_2 or too-high CO_2 concentrations.

Use of CA technology is limited to relatively few FFV, the major crop being apples. Other FFV include cabbages, sweet onions, kiwifruits, avocados, persimmons, pomegranates, nuts and dried fruits, and vegetables. Atmospheric modification during long-distance transport is used on apples, asparagus, avocados, bananas, broccoli, cranberries, cherries, figs, kiwifruits, mangos, melons, nectarines, peaches, pears, plums and strawberries. Limited use of CA for many FFV is related to the high levels of capital investment required for high-quality storage rooms and the maintenance and monitoring of atmospheres; therefore, fruit volumes must be high enough to fill rooms, and extended storage periods are needed to make investments economical. For shipping containers, limitations include maintaining container identity for those with equipment. Specific CA storage technologies that are used are a function of the FFV, growing region and size of the industry. Where used, static systems are still used for the majority of FFV CA storage, ranging from simple controlled ventilation systems to conventional CA (more than 2 kPa O_2). The use of low O_2 (LO) and ultralow O_2 (ULO) CA storage, where O_2 levels are as low as 1.5–2.0 and 0.8–1.2 kPa, respectively, is increasingly becoming common. Chong et al. proposed a hollow fibre membrane module for generating nitrogen-enriched air stream that controls the concentration of both O_2 and CO_2 in the CA storage.

Other developments include dynamic CA (DCA), initial low O_2 stress (ILOS) and hypobaric storage. The principle of DCA is that responses of fruit to decreasing O_2 in the storage atmosphere are monitored. As O_2 levels reach the anaerobic compensation point, stress responses by the fruit can be detected and the O_2increased to slightly higher levels. The most common commercial DCA system is based on chlorophyll fluorescence (DCA-DF;), but others include measurement of ethanol in the fruit (dynamic controlled systems, DCS;) and the respiration quotient (DCS-RQ;). Most research has been carried out with apples and pears, a major impetus for research being the loss of diphenylamine, the antioxidant used widely for control of the physiological disorder superficial scald in Europe. However, DCA has been tested with avocado and should work with any chlorophyll-containing product. ILOS involves the use of low partial pressures of O_2, e.g. 0.4 kPa, that result in ethanol accumulation for short periods of time that result in delay of fruit ripening. Hypobaric or low-pressure storage results in very low O_2 around the product and delay ripening and senescence but has yet to find commercial acceptance because of factors such as cost and safety. These issues may be addressed in the future.

Technologies

Plasma is an emerging technique for decontaminating FFV. Plasma is composed of ionized gas molecules, which have been dissociated via an energy input. Depending on the mode of particles activation and the excitation energy, they can generate high or low temperatures, referred to as thermal or cold plasma, respectively. Cold plasma at atmospheric pressure can be generated by transforming argon gas into plasma at radio frequency of 27 MHz or by electric discharge between two electrodes separated by dielectric barriers.

Three basic mechanisms have been suggested for the inactivation microbial spores in plasma environments, including the erosion of microbial spore surface atom by atom through adsorption of reactive free radicals ‘etching’; direct destruction of DNA via UV irradiation and volatilization of compounds from the spore surface by UV photons through intrinsic photo-desorption. Fernández et al. revealed that at the optimal operating conditions of cold gas plasma treatment about 15 min treatment times was required to achieve 2.72, 1.76 and 0.94 log-reductions in viable cells of S. enteric sv. Typhimurium on lettuce, strawberry surfaces and potato tissue respectively. Recent study on fresh corn salad leaves showed that the plasma treatment at 20 W for 1 min successfully inactivated E. coli by 4 log-cycles. However, more research is required for a complete understanding of the role of microbial cell structure, physiology and stress resistance mechanisms involved in plasma resistance. Also, the effect of plasma treatment on food enzymes and postharvest quality attributes of FFV requires more detailed study. Safety of gases, consumer perception and the translation of laboratory scale to large commercial scale, also requires further investigation.

POSTHARVEST COOLING AND HANDLING METHODS

Producing the consistently high-quality produce that spells success in a highly competitive market does not stop at the end of the row. The ability to deliver a quality product to the market and ultimately to the consumer commands buyer attention and gives the grower a competitive edge. Proper postharvest cooling and handling can help to ensure that quality is maintained until the product reaches the consumer.

Postharvest cooling rapidly removes field heat from freshly harvested commodities before shipment, storage, or processing and is essential for many perishable crops. Proper postharvest cooling can:

- Suppress enzymatic degradation and respiratory activity (softening).
- Slow or inhibit water loss (wilting).
- Slow or inhibit the growth of decay-producing microorganisms (molds and bacteria).
- Reduce production of ethylene (a ripening agent) or minimize the product's reaction to ethylene.

In addition to protecting quality, postharvest cooling provides marketing flexibility by making it possible to market at the optimum time. Being able to cool and store produce eliminates the need to market immediately after harvest, which can be an advantage for high-volume growers as well as pick-your-own (P-Y-O) operators who wish to supply restaurants and grocery stores. To select the best cooling method, it is necessary to understand the basic principles of cooling. The choice of cooling method depends on the following factors:

- The nature of the product: Different types of produce have different cooling requirements. For example, strawberries and broccoli require near-freezing temperatures, whereas summer squash or tomatoes would be damaged by such low temperatures. Likewise, because of problems that can be caused by wetting of certain products, hydrocooling or icing may not be appropriate.
- Product packaging requirements: The best choice of cooling method may depend on whether the produce is in a box, bin, or bag. The package design can have an effect on the method and rate of cooling.

- Product flow capacity: Some methods of cooling are much faster than others. If the volume of produce to be cooled per season, per day, or per hour is large, it may be necessary to use a faster cooling method than would be used for lower volumes.
- Economic constraints: Construction and operating costs vary among cooling methods. The expense of cooling must be justified by higher selling prices and other economic benefits. In some cases – for example, when the volume of produce is low – the more expensive methods cannot be made to pay for themselves.

Common Produce Cooling Methods

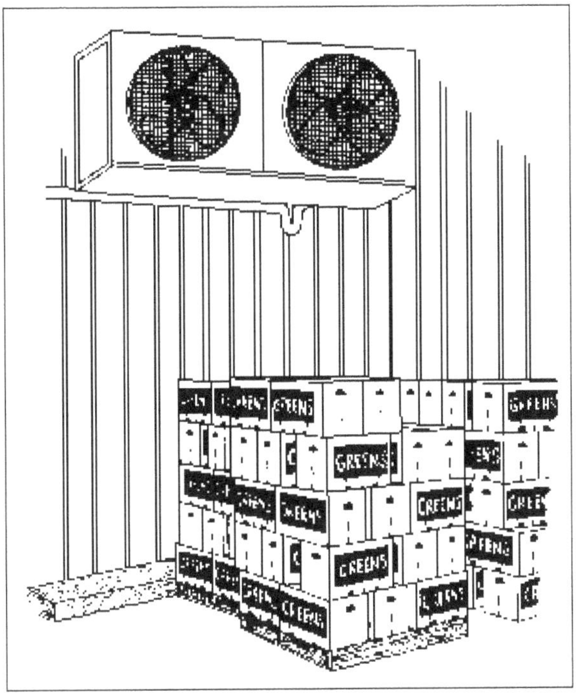

Room cooling.

Room cooling is simply a matter of placing produce in an insulated room equipped with refrigeration units to chill the air. It may be used with most commodities but may be too slow for some that require quick cooling. It is effective for storing precooled produce but in some cases cannot remove field heat rapidly enough. Carefully directing the output of the cooling system evaporator fans can significantly improve the cooling rate. Properly designed, a room cooling system can be relatively energy efficient.

A cooling room that is used only to store previously cooled produce requires a relatively small refrigeration unit. However, if it is used to cool the produce, a larger unit is needed. To determine the correct size of the refrigeration unit, consult a refrigeration contractor or Cooperative Extension.

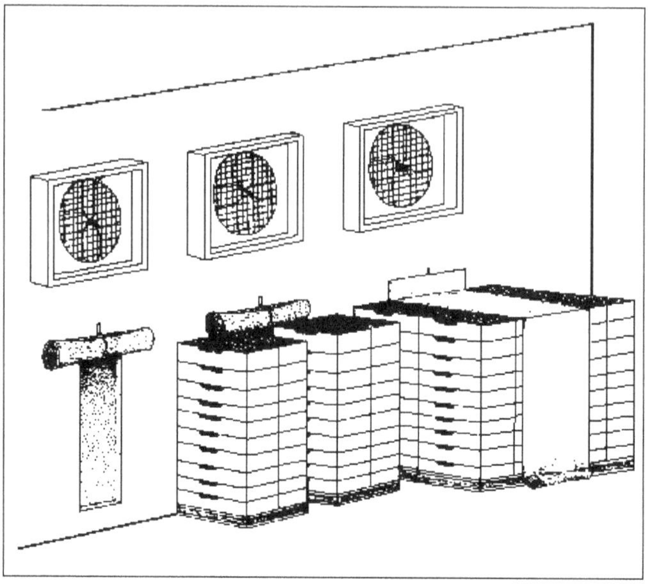

Forced-air cooling.

Forced-air cooling is used in conjunction with a cooling room and can be used effectively on most packaged produce. To increase the cooling rate, additional fans are used to pull cool air through the packages of produce. Although the cooling rate depends on the air temperature and the rate of airflow through the packages, this method is usually 75 to 90 percent faster than room cooling.

When forced-air cooling fans are added to an existing cooling room, it may be necessary to increase the size of the refrigeration unit to accommodate the additional cooling load. The fans are normally equipped with a thermostat that automatically shuts them off as soon as the desired produce temperature is reached to reduce energy consumption and water loss from the produce. Forced-air cooling can be very energy efficient and is an effective way to increase the heat removal rate of a cooling room.

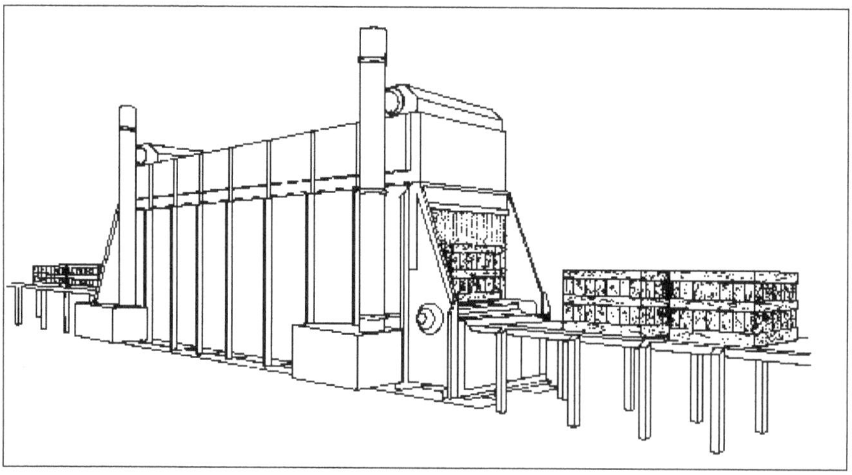

Hydrocooler.

Hydrocooling can be used on most commodities that are not sensitive to wetting. (Wetting often encourages the growth of microorganisms). In this process, chilled water flows over the produce, rapidly removing heat. At typical flow rates and temperature differences, water removes heat about 15 times faster than air. However, hydrocooling is only about 20 to 40 percent energy efficient, as compared to 70 or 80 percent for room and forced-air cooling.

Hydrocoolers were once rare but are becoming more common because they are able to cool a large amount of produce rapidly.

Top/liquid icing.

Top or liquid icing may be used on a variety of commodities. In the top icing process, crushed ice is added to the container over the top of the produce by hand or machine. For liquid icing, a slurry of water and ice is injected into produce packages through vents or handholds without depalletizing the packages or removing their tops.

Icing is particularly effective on dense packages that cannot be cooled with forced air. Because the ice has a residual effect, this method works well with commodities that have a high respiration rate, such as sweet corn and broccoli. Icing is relatively energy efficient. One pound of ice will cool about 3 pounds of produce from 85 to 40 °F.

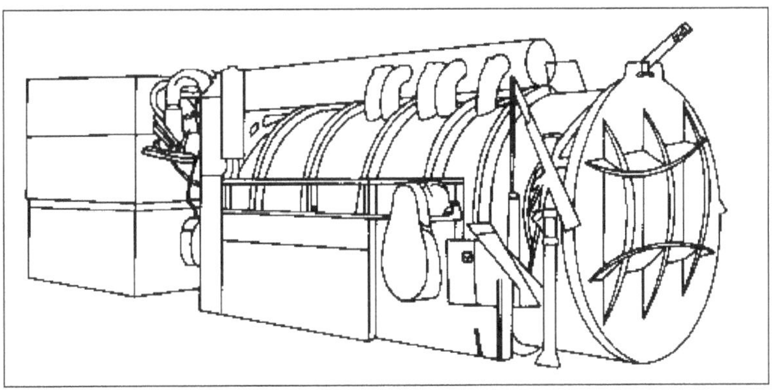

Vacuum cooling.

Vacuum cooling is effective on products that have a high ratio of surface area to volume, such as leafy greens and lettuce, that would be very difficult to cool with forced air or hydrocooling. The produce is placed inside a large metal cylinder and much of the air is evacuated. The vacuum causes water to evaporate rapidly from the surface of the produce, lowering its temperature. The process may cause wilting from water loss if overdone. Vacuum coolers can be energy efficient but are expensive to purchase and operate. They have only limited application to most common types of produce.

Evaporative cooling is an effective and inexpensive means of providing a lower temperature atmosphere with high relative humidity for cooling produce. It is accomplished by misting or wetting the produce in the presence of a stream of dry air. Evaporative cooling works best when the relative humidity of the air is below 65 percent. At best, however, it reduces the temperature of the produce only 10 to 15 °F and does not provide consistent and thorough cooling.

Produce should be picked during the coolest parts of the day and kept in the shade away from direct sunlight. In extremely hot weather, many growers in the western states have found that harvesting at night, "under the lights," is a good way to reduce the amount of mechanical cooling needed.

Other Steps for Maintaining Quality

Refrigerated transportation vehicles are not designed to remove field heat from produce. At best they can only maintain the temperature at which the produce was loaded. Growers, shippers, and truckers can contribute to better produce temperatures and hence fewer losses during transport by following these guidelines:

- Precool trailers before loading them, especially in warm weather.
- Measure and record the product's temperature during loading.
- Keep the trailer and refrigeration unit in good repair.
- Load the produce in a way that will ensure maximum air circulation.
- Keep transit times to a minimum.

ULTRAVIOLET LIGHT TREATMENT

Fruits and vegetables are highly perishable due to their susceptibility to fungal infection, tissue senescence and transpiration. As a result, the horticulture industry is constantly seeking safer and more effective preservation techniques for fresh produce in order to extend shelf-life, maintain resistance to diseases and retain beneficial phytochemicals.

Ultraviolet (UV) treatment of fresh produce is considered an alternative to chemical approaches that has great potential for controlling postharvest diseases. Although UV light is generally harmful to living systems at high doses, at low doses it induces disease resistance in horticultural crops, slows ripening and maturation rate, and improves customary quality attributes. In specific cases, UV light also improves phytochemical content. This biological phenomenon is known as hormesis, where a low dose of a harmful physical or chemical agent stimulates beneficial responses in biological systems. The beneficial doses of the stressors are termed hormetic or hormic doses.

UV Hormesis in Horticultural Crops

Harvested plant organs intended to be stored for subsequent consumption in the fresh state have various morphogenetic origins. They also differ in their postharvest behaviour and in their ability to respond to different postharvest practices and treatments. Differences in sensitivity to UV light also exist among species in field-grown plants. Therefore, it is not surprising that postharvest plant organs react differently to different doses of UV-C applied to prolong their postharvest life. Studies by several research groups have shown the existence of a defined level of UV treatment that stimulates beneficial responses in a variety of crops. The reported hormetic UV doses for treatment of fruits and vegetables range from 0.125 to 9 kJ/m^2. The lowest and the highest hormetic doses were reported for table grape and orange, respectively. For most commodities, a single dose is characterised as hormetic, although for some a range of doses is considered beneficial. The dose required for a specific response also differs between crops. For example, the hormetic dose required for controlling a fungus in apple was different from the dose required in grapefruit. In addition, seasonal differences (eg, for grapefruit) and varietal differences (eg, for sweet potato) affected hormetic doses. In some instances, only a minor difference in the hormetic dose requirement was found by two independent groups (eg, in mature green tomato).

Disease Control

The Phenomenon of Disease Resistance

Effective control of several common pathogens that cause decay of postharvest crops was achieved by pre-storage treatment with defined hormetic UV doses. Both fungal and bacterial rots were reduced by pre-storage application of the hormetic UV treatment. Several factors can affect disease control in UV-treated commodities. Different hormetic doses were necessary for controlling different pathogens on the same produce. Maturity and physiological age also affected the responsiveness of the treated tissue. Tomato fruits irradiated at the mature-green stage showed higher resistance than fruits treated at the breaker stage. The level of resistance induced by UV-C was also attenuated in older carrot roots compared with fresh roots. Varietal differences were also observed, who reported that the hormetic dose for the sweet potato cultivars Jewel and Carver was 4.8 kJ/m^2 versus 3.6 kJ/m^2 for Georgia Jet. A higher dose was required for maximum resistance against Penicillium digitatum as the season progressed. The development of disease resistance appeared to be a gradual process. The latter two studies reported that two citrus fruits (lemon and grapefruit) were as sensitive to green mould as the control when inoculation was performed immediately or a few hours after UV treatment. The maximum level of resistance in citrus fruits was observed 48 h after treatment. In carrot slices, the establishment of the induced resistance proceeded at a slower rate and was observed only after 7 days, presumably because of the lower storage temperature. The latency period was even longer in whole carrot roots, in which the resistance response to Botrytis cinerea was observed only after 25 days. In tomato fruits, resistance to B. cinerea was evident within 72 h after UV treatment, preceded by a period of high sensitivity compared with the control. Expression of the induced disease resistance was also affected by storage temperature. Lower temperatures, 1 °C versus 4 °C in carrot slices and 6 °C versus 11, 17, 20 or 25 °C in grapefruit, were more effective in maintaining the resistance response.

The Mechanisms of Disease Resistance

The observed reduction of disease incidence in UV-C-treated fruits and vegetables does not for the most part seem to be due to surface sterilisation of the tissue or a germicidal effect on the pathogens, although some decontamination could occur. The maintenance of the juvenile state may explain, in part, the resistance of UV-treated tissue to pathogenic fungi.

It is likely that UV treatment triggers more than one defence mechanism, but there is more information available on the enhancement of preformed antifungal compounds and on the elicitation of phytoalexins, which are low-molecular-weight secondary metabolites with antibacterial and antifungal activities. The resistance of immature tomato fruits to B. cinerea is attributed to the presence of the glycoalkaloid tomatine. In UV-treated tomato, higher levels of tomatine were detected over a storage period of 96 h after treatment. A low dose of UV-C was shown to induce various classes of phytoalexins in several postharvest commodities. Several studies present solid evidence of a positive correlation between phytoalexin accumulation and the observed disease resistance. In carrot slices and roots, the accumulation of the isocoumarin phytoalexin, 6-methoxymellein (6-MM), was strongly correlated with the development of resistance to B. cinerea and Sclerotinia sclerotiorum. In citrus fruits (grapefruit, kumquat, lemon, orange) resistance to Penicillium digitatum and Penicillium italicum was linked to the accumulation of scoparone or scopoletin. The expression of resistance in tomato fruits to B. cinerea was concomitant with induction of the phytoalexin rishitin in the UV-treated tissue. The phytoalexins isalexin, S-(-)- spirobrassinin, 1-methoxybrassitin, brassicanal C and caulilexins A–C were detected in cauliflower subjected to a low dose of UV-C. The antifungal activities of caulilexin A, caulilexin C and brassicanal C, three phytoalexins induced by an unspecified UV-C dose, were evaluated in bioassays with Leptosphaeria maculans, Rhizoctonia solani and Sclerotinia sclerotiorum. From these bioassays, caulilexin A was characterised as the metabolite with the highest antifungal activity against R. solani and S. sclerotiorum. Resistance of table grape to B. cinerea and other pathogens both during growth and at the postharvest stage has long being associated with the UV-C–inducible phytoalexins resveratrol, ε-viniferin and α-viniferin.

The phytoalexin response in UV-treated commodities is modulated by the physiological status of the plant organ at the time of the treatment, as well as by storage duration. Wounded carrot roots were shown to accumulate higher levels of 6-MM in response to UV treatment than unwounded roots. It was also observed that the physiological age of carrot root and tomato fruits affected the intensity of the phytoalexin response. Roots treated immediately after harvest (fresh roots) accumulated higher levels of 6-MM than older roots (stored for 4 months). Older carrot roots also required a higher dose to express the phytoalexin response. When treated at the mature-green stage, tomato fruits showed a greater ability to synthesise and accumulate rishitin than fruits at the turning stage. A similar age effect was also reported for orange. Storage conditions, namely temperature and relative humidity (RH), had a significant impact on the phytoalexin response. Tomato fruits stored at 13 °C and 95% RH for 12 or 22 days had a higher rishitin content than fruits stored at 20 °C and 75% RH. A higher level of 6- MM was found in carrot slices stored at 4 °C than in those stored at 10 °C.

Varietal differences also exist in the ability to accumulate phytoalexins in response to UV. This was demonstrated with carrot roots in the accumulation of 6-MM. Related species of a single family also vary in their capacity to accumulate the same phytoalexin in response to UV. Significantly higher levels of scoparone accumulated in lemon, kumquat and orange, but accumulation was low in grapefruit and lime. It was also observed that heat treatment combined with inoculation (P. digitatum) induced a higher level of the phytoalexin in these fruits than UV-C treatment alone Furthermore, the phytoalexin profile may be altered depending on the UV dose. In kumquat fruits treated with a low UV dose (0.3 kJ/m^2), scopoletin was the major phytoalexin, whereas at a higher dose of 1.5 kJ/m^2, scoparone was the main phytoalexin isolated from the fruit flavedo.

Defence mechanisms other than phytoalexin response have received little attention thus far, although the potential for the expression of other defence-related genes in postharvest commodities exists. A study presented evidence for the involvement of pathogenesis-related (PR) proteins, namely chitinase and β-1,3-glucanase, in the resistance of UV-irradiated grapefruit to P. digitatum. While UV treatment alone can cause chitinase expression, a combination of both wounding and UV treatment was necessary to trigger β1,3-glucanases in grapefruit flavedo tissue. Antifungal hydrolases, some of which are PR proteins, were reported in bell pepper, tomato and carrot roots in response to UV treatment. In carrot roots, a chitinase was induced by UV-C application but also by infection with B. cinerea, a finding that confirms the PR status of this hydrolase. Both chitinases and β-1,3-glucanases were induced in tomato fruits by UV-C alone. Increased chitinase and β-1,3-glucanase activities were detected in peach fruit extracts as early as 6 h after UV treatment, reaching a maximum level at 96 h. Enhancement of both chitinase and glucanase activities were also observed in UV-treated grape leaves.

UV irradiation is known to stimulate the phenylpropanoid pathway of all plant parts, as demonstrated by the activation of the key enzyme of this pathway, phenylalanine ammonialyase (PAL). PAL activity was enhanced in UV-treated grapefruit, sweet potato and peach. The metabolites from the phenylpropanoid pathway exert a protective role against pathogenic organisms through direct inhibition of growth, inactivation of fungal hydrolases or reinforcement of host cell walls. Elevated phenolic content was noticed in UVtreated strawberry and tomato, likely contributing to the inactivation of fungal hydrolases. Histochemical and ultrastructural studies provide evidence of the possible involvement of phenolic compounds in resistance to B. cinerea. Cell-wall-strengthening compounds (lignin and suberin) were involved in the observed resistance of UVtreated tomato fruits to B. cinerea. The resistance of UV-treated grapefruit to P. digitatum was associated with enhanced activity of peroxidase, a key enzyme involved in lignin synthesis.

Maintenance and Improvement of Quality Attributes

Control of Senescence

The rate of metabolic activities in UV-C–treated tissue was modified following treatment as indicated by the change in the respiratory pattern reported for some postharvest commodities. In tomato fruits a rapid and transient increase in CO_2 production immediately after irradiation was reported. Compared with the control, however, the overall CO_2 production was lower in the UV-treated fruits, and the CO_2 climacteric rise was delayed by 7 days. Ethylene (C_2H_4) synthesis followed a similar pattern. The delayed climacteric rise (CO_2 and C_2H_4) was interpreted as indicative of delayed senescence resulting from UV treatment. The transient rise in CO_2 and C_2H_4 observed in tomato following UV treatment was presumed to be an adaptation of the tissue to the oxidative stress caused by UV light. The data obtained indicate that the stimulation of multifunctional antioxidants such as polyamines (namely putrescine, agmatine and tyramine) and phenolics (namely salicylic acid, vanillin, ferulic acid, 7-hydroxycoumarin and rutin) may account for the observed delay in senescence. Detoxification of active oxygen species may also play a role in retarding senescence.

Changes in Firmness

Firmness is a very important quality factor for postharvest fruits and vegetables. Firmer produce are better able to withstand postharvest handling and transportation. Nevertheless, tissue softening is essential for most fruits towards the end of storage to impart desirable palatability. UV-C irradiation appears to slow down the softening rate of several fleshy fruits. Higher firmness indices were reported for UV-treated strawberry, peach, apple and pepper. Similar results were observed for UV-treated tomato fruits, which remained firmer over a longer period compared with the control. In UV-treated tomato, the activity of polygalacturonase, an enzyme associated with fruit softening during ripening, was lower than in the control. The activity of other cell-wall-degrading enzymes was also found to be reduced in UV-treated tomato. However, a sensory panel did not find a significant difference in texture between UV-treated tomato and the control at the end of the storage period.

Changes in Taste and Flavor Attributes

The effect of UV-C treatment on attributes contributing to taste, such as free sugars assessed as total soluble solids and titratable acidity, have been studied in several UV-treated horticultural crops. Baka found a higher level of free sugars in treated fruits than in controls; strawberry and bell pepper. In UV-treated peach, apple, onion and sweet potato, however, the sugar content was lower, although in the case of onion, the difference was not significant. Charles et al. did not observe any significant difference in the evolution of total soluble solids in UV-treated tomato during storage, nor did Vicente et al. in UV-treated pepper.

Titratable acidity was found to decrease at a slower rate in UV-treated strawberry and bell pepper, but UV-C did not significantly affect either the pH or the acidity of apple. While the latter study reported a similar pattern for the pH value of peach, the acidity was significantly higher in the treated fruits. An opposite trend in titratable acidity was noticed in UV-treated tomato fruits during 20 days of storage at 13 °C, although observed differences were not always statistically significant.

Phenolic compounds can contribute to the flavour of fruits. Maharaj reported significant increases in phenolic compounds in UV-treated tomato fruits. However, a sensory panel found UV-treated tomato fruits to be comparable in taste to either control or store-bought fruits. Similarly, UV-treated citrus fruits (kumquat and Washington navel orange) did not differ in taste from the control. The accumulation of the isocoumarin 6-MM is associated with the development of a bitter taste in stored carrot roots exposed to ethylene. Although a significant level of 6-MM was induced in carrot roots and slices in response to pre-storage UV-C treatment, sensory evaluation of the cooked carrot roots showed no significant difference between the treated roots and the control. It was shown that 6-MM leached out into the boiling water and that the peeling operation also reduced the 6-MM level in the cooked slices, since this compound accumulates mainly in the epidermic and exodermic tissues. In addition, a sensory panel could not differentiate UV-treated onions from the control, both in the fresh and cooked forms.

Changes in Color and General Appearance

The colour and general appearance (turgidity or shrivelling) of fresh stored fruits and vegetables can be a decisive factor in their acceptance or rejection by the consumer. During ripening, fruit colour change is due to chlorophyll loss, the unveiling of pre-existing carotenoids, and the synthesis of carotenoids such as lycopene and other pigments such as anthocyanins. From the standpoint of prolonging shelf-life, it is desirable that these changes take place at a slower rate. Several authors have described the delay in colour change induced by UV-C in treated fruits and vegetables. In pepper and broccoli and tomato, UV-C irradiation significantly delayed chlorophyll loss. Concomitant with this delay, the evolution of lycopene was also slowed in tomato. However, colour development was accentuated by UV treatment in strawberry fruits with a concomitant increase in anthocyanin accumulation.

Doses higher than the hormetic dose led to detrimental bronzing and duller appearance in tomato fruits and altered the appearance of citrus fruits. Fruit surface scalding occurred in papaya at all UV doses applied. At higher doses, yellow lemon fruits exhibited less damage than green fruits. The negative effect of UV-C treatment on appearance could lead to rejection of this promising alternative, at least for some crops (eg, avena mandarins). D'hallewin et al. did not consider UV-C to be an effective method compared with either heat treatment or thiabendazole, mainly because of the poor visual appearance of the UV-C-treated fruits, even though the level of decay control was significant (less than 50% of the control). The consumer usually judges food products as sound when they are free of any visual defects. This may raise the question of choice for the consumer between reduced chemical residues in food and a lower esthetic value of the UV-C-treated fruits or vegetables. It is possible that the extensive loss of sensory quality observed by D'hallewin et al. might have been due to overexposure. Similar quality losses were also observed in other crops when doses were higher than the hormetic dose.

Weight loss in postharvest crops during storage may be attributed to loss of water through transpiration or to loss of substrates (sugar and organic acids) through respiration. Water loss may cause shrivelling of the stored commodity, with a serious effect on general appearance. In bell pepper, a combination of UV treatment and hot water dipping resulted in lower weight loss than in the control. A modest weight loss was also noted in UV-treated sweet potatoes, but weight loss was higher in UV-treated tomato fruits than in the untreated control, although not significant enough to cause shrivelling. UV-C treatment caused severe shrivelling of kumquat when the produce was stored at 17 °C. This deleterious effect, however, was reduced by storage at lower temperatures.

Changes in Bioactive Compounds

UV hormesis is seen as a technology that could also be used to enhance the beneficial health properties of fresh fruits and vegetables. This hypothetical view is based on the ability of this abiotic elicitor to activate several paths leading to the synthesis and accumulation of secondary metabolites, some of which are well characterised for their positive impact on several aspects of human health. The most cited example is the elicitation of the stilbene resveratrol in table grape by UV treatment. This molecule has been widely studied for its potential to protect against or alleviate the symptoms of several human ailments and diseases. The prophylactic and therapeutic actions of resveratrol have been described as antioxidant, cancer chemopreventive, anti-infection, antiviral, anti-inflammatory, anti-ischaemia, cardioprotective, neuroprotective, anti-aging and obesity-controlling. UV treatment increased the level of resveratrol 10-fold in table grape skin compared with untreated fruit skin. Besides table grape, numerous plants have been identified as resveratrol sources. There is therefore a potential for the biosynthetic pathways leading to increased resveratrol accumulation to be activated by UV hormesis in other edible plant foods.

The bioactive coumarin scoparone, naturally isolated from the Chinese herb Artemisia scoparia, was induced in citrus fruits by UV hormesis, as well as by heat treatment. Scoparone was shown to be an antiproliferative agent and a smooth muscle relaxant, and to reduce total serum cholesterol and triglycerides. In a hyperlipidaemic diabetic rabbit model, scoparone was shown to lower plasma cholesterol and slowed the development of atherosclerosis. However, a preliminary study using a murine model was not conclusive with respect to the beneficial effect of scoparone-enhanced UV-treated kumquat fruits peel. Animals that were fed freeze-dried peels with an alleged 30-fold increased level of scoparone did not show any hypolipidaemia compared with control animals fed nonUV-treated kumquat peels, presumably because of a reduction in the total antioxidant levels in the peel of the UV-treated fruits.

Degenerative, chronic and age-related diseases all have an oxidative component, and increased consumption of foods with a high antioxidant potential is recommended for prevention of these diseases. Fresh fruits and vegetables are an integral part of a healthy diet due to their high antioxidant potential, often linked to their antioxidant vitamins, phenolic compounds and pigments. UV-C illumination of fresh strawberry fruits was shown to cause a significant enhancement of the total antioxidant capacity. The increased antioxidant capacity of UV-treated strawberry was likely due to the elevated levels of phenolic compounds, although the titre of anthocyanin pigments was lower than in the control. The results of Erkan on the effect of UV on anthocyanins in strawberry contradict the results. Higher anthocyanin content was noted in the treated berries at the end of the storage period. In other fruit systems, such as ready-to-eat pomegranate arils and table grape, UV-C treatment did not cause any significant change in anthocyanin content.

At least one study reported a change in the carotenoid profile in UV-treated tomato. Reductions in lycopene content in UV-treated tomato fruits and in total carotenoids in pepper were observed. The negative impact of UV-C treatment on carotenoids may be considered a drawback, particularly with regards to lycopene, which is highly valued as a phytochemical that is beneficial to human health. On the other hand, both histochemical and chemical analyses revealed higher phenolic content in UV-treated tomato. Similarly, UV-C treatment resulted in a significant increase in total phenolics in strawberry fruits. The increase in phenolic compounds in UV-treated fruits may well compensate for the reduction in antioxidant pigments, resulting in a higher antioxidant capacity of UV-treated fruits as in strawberry. The antioxidant capacity was also higher in UV-treated pepper.

DRYING

Drying is an important step in post-harvest technology because every agricultural product in its natural state contains a certain amount of water, called the moisture percentage. In order to preserve the quality of the product and its storage life, this moisture content needs to be reduced to a level that is safe for storage. Drying is, therefore, a key process for managing the product moisture.

Any crop type (grains, vegetables, fruits, herbs) can undergo the process of drying.

Although drying is one of the oldest methods of food preservation, this is a complex operation which involves heat application over the product. Consequently, the product may change its color, texture, and physical characteristics.

Drying Process and Conditions

Drying preserves products by removing enough moisture to prevent decay and spoilage. The water content of properly dried products varies from 5 to 25%, depending on the product type. Moreover, successful drying depends on:

- High temperature needed to draw out moisture.
- Dry air required to absorb the released moisture.
- Adequate air circulation to carry out the moisture.

Airflow rate, air temperature, and relative air humidity all together influence drying speed. In general, higher airflow rates, higher air temperatures and lower relative humidity increase drying speed.

Drying speed is also influenced by the content of moisture in the product. For instance, it takes much longer to lower the content of moisture in a product from 25 to 15 percent, than from 35 to 25 percent. This is due to the fact that water present in the outer layers of a product evaporates much faster and more easily than the one in the internal layers. One common misconception is that rapid drying at high temperatures can overcome this problem. In fact, such drying conditions create internal tensions, producing tiny cracks that can lead to ruptures of the product (grains) during the treatment.

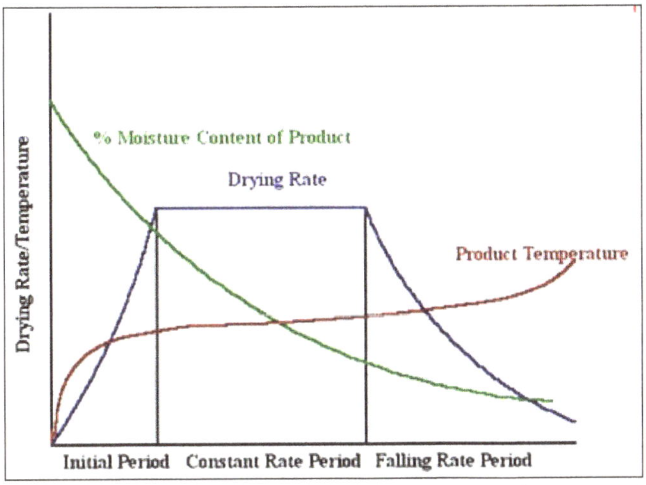

Successful drying is closely related to the content of moisture in the product, drying temperature, and air conditions.

In the drying process, the key is to remove moisture from the product as fast as possible, at a temperature that does not seriously affect the flavor, texture, and color. Too low temperature at the beginning can cause the growth of microorganisms, thus affecting its quality. On the other hand, too high temperature in a combination with too low humidity can harden the product's surface, making it more difficult to dry out. Finally, delays in drying, incomplete or ineffective drying will reduce product quality and result in losses.

Drying is a very simple, ancient skill. It is one of the most accessible and hence the most widespread processing technology. Here are two common methods used in drying of various products:

- Natural drying; such as sun drying.
- Artificial drying; drying with low (aeration) and high temperatures (heated air drying, dryeration, and in-storage cooling).

Sun drying of cranberries

Smallholder farmers usually dry their products naturally; with or without shade.

In natural drying, the product (mainly grains) is spread on the ground, in sun or shade, for a short period. To provide unified drying, the product needs to be regularly stirred. Although this method is simple, it has some limitations–slow drying can harm the product's quality, therefore it's not recommended in humid areas due to excessive moisture, and increased relative air humidity at night can speed up dehumidification of the product.

Artificial drying provides faster indoor drying in a controlled environment. The product is exposed to forced ventilation of air, which is heated to a certain degree in special appliances called dryers. Artificial drying is a great solution for faster drying of high-value products as well as preserved product quality. However, purchasing of a dryer, fuel, and other elements necessary for proper work, makes this type of drying expensive and less available to low-income farmers.

Artificial drying includes indoor drying where several conditions may be achieved; drying of various grains.

As stated, the choice of which drying system to choose depends mainly on the type of crop the farmer grows and his financial ability to afford it.

Classification Criteria of Crop Dryers

Today, there are over 500 different dryer types known. They all differ on crop type, mode of operation, operating pressure, state of the material, heat input, drying medium as well as on drying temperature.

Table: Types of dryers based on a certain criterion.

Criterion	Type
Mode of Operation	Batch, continuous
Operating Pressure	Vacuum, atmospheric
State of Material	Stationary, moving
Heat Input	Convection (direct), conduction (indirect), radiation electromagnetic fields(MW,RF), and combined mode

Drying Medium	Air, superheated steam, flue gases
Drying Temperature	Below and above boiling temperature, below freezing point

From a scientific point of view, drying is a process of liquid vaporization due to heat application onto the wet product.

Modern facility for grain drying-silos.

Therefore, the byproduct of every drying process is water. The product releases water in two phases. In the first phase, it releases intra-cellular water with nutrients and dissolved solids also called "free water". As the process continues, the product releases "bound water". This water is built into cells or biopolymer structures and therefore it needs additional energy to break the bonds with solids. For safe storage, products must be dried at the appropriate levels and stored under adequate conditions.

SORTING AND GRADING

Separating the different types of fruits from fruit lot is called sorting, it depends on:

- Damage product: Damaged product should be avoided. It should be separated from fruit lot and well mature product should be taken.
- Diseases: Diseased free fruit and vegetable should be collected from a lot. It is very important for sorting.
- Insect cutting products: Insect cutting fruit must be avoided because consumers want good fruit.
- Maturing: Mature fruit is very tasty. So maturity is very much needed for fruit.

- Color: Colours should be lightable. Well coloured fruit should be selected that attract consumer. Colour also indicates the maturity level of seed.
- Shape: In case of packaging, shape is essential. Some type of fruit should be selected for packaging in case of sorting.
- Size: Size also an important matter for sorting fruits from a fruit lot. We should selected same size of fruit.

Grading

Grading of fruits and vegetables after harvesting is an essential step in post-harvest management. Grading of fruits and vegetables on the basis of physical characteristics like weight, size, colour, shape, specific gravity, and freedom from diseases depending upon agrocliamatic conditions. The known methods of grading of fruits and vegetables are manual grading, size grading. Grading of fruits and vegetables in the fresh form for quality is essential, as the people are becoming quality conscious day by day. Further, upon arrival of fruits and vegetables at the processing centres, they should be graded strictly for quality. The immature, properly mature and over mature fruits and vegetable should be sorted out for the best attributes.

Grading is sorting of vegetables and fruits into different grades according to the size, shape, colour, and volume to fetch high price in market or, Arrangement of fruits into different groups by separating from a fruit lot is called grading. Grading depends on:

- Cultivar.
- Size.
- Appearance.
- Colour.
- Quality.

Objectives

- To get higher price.
- To have different marketing value.
- To adjust with world market.
- To facilitate marketing.
- To facilitate packing.
- To facilitate transporting.
- To increase the shelf life.

For International market, three general grades are considered as:

- Extra Class: The extra class is of superior quality posses the shapes and colour of the variety and without internal defect likely to affect the inherent texture and flavour. A 5% tolerance is allowed for errors. It must be carefully presented taking into accounts the uniformity of the produces in size colour, condition arrangement of the produce in the package quality and appearances of the packing or pre-packing material.
- Class I: Almost having a same quality is like the Extra Class except that a 10% tolerance is allowed. Individual fruit is allowed a slight defect in shape, colour and miner skin defect which do not affect the general appearance for keeping qualities. In packing the size range may be wider and product need not always be arranged in the package.
- Class II: This class product may exhibit some external or internal defects provided they are fit for consumption while fresh. This class is best fitted for local or short distance market. This category will satisfy the needs of customers who are not too demanding and for whom price is more important than quality.

Advantages of Grading

- Losses the selling price due to presence of substandard products or specimen can be easily avoided.
- It increased marketing efficiency by facilitating buying and selling a produce without personal selection.
- Grading enhanced to set good price for graded products.
- Heavy marketing cost in packing and transportation can be avoided by grading.
- In grading, diseased and defected specimen are not damaged due to contact of diseased specimens and thus gets high price in market.
- By grading, there is fairness to both Buyers and Sellers.
- Properly graded vegetables and fruits are purchased by the consumer easily without inspection.

Grading of Fruits

Generally, the fruits are graded on the basis of size, weight, specific gravity, colour, variety, etc. Size grading is predominantly followed in almost all types of fruits on the basis of size. The fruits are graded as a small, medium, large and extralarge. On the basis of maturity, the fruits are graded as immature, properly mature and over mature.

Grading on the basis of maturity decides both quality and shelf life. The Alphonsoad and Pairi mango fruits are graded on the basis of weight as less than 200g, 200-249g, 250-299g, 300-349g and more than 350g. Out of these grades the weight grade 250-299g account for about 30% of the fruits. The mango fruits are also graded on the basis of Sp. gravity (3 grades on the basis of sp. Gravity as less than 1.0 sp.gr, 1.0-1.02 and more than 1.02). The sp.gr Grade 1.0-1.02 accounts for about 50% of the Alphonso and Pairi mango fruits.

Grading of Vegetable

The fruit vegetables such as bitter gourd, okra, bell pepper, brinjal, green chilli etc. also graded on the basis of size into 3 grades as small, medium and large. The vegetables like tomato are graded on the basis of colour.

PRE-COOLING

Pre-cooling is essential for maintaining an optimal postharvest life and to minimize the impact of the product's senescence on its market quality. Pre-cooling (also called "preliminary cooling") is a vital post-harvest management practice that is required prior to storage.

Since there is a big difference between the temperature of harvested crops and cooling storage, the main purpose of pre-cooling is to remove the heat from the crops after they are harvested. That way, pre-cooling reduces fruit respiration rate and extends crop storage life. Along with that, pre-cooling also inhibits the growth of various microorganisms, thus reducing the possibility of post-harvest diseases.

The success of pre-cooling will depend upon two important factors:

- The time between the harvest and pre-cooling.
- Initial crop temperature before pre-cooling.

Generally, it's recommended to practice pre-cooling as soon as possible after the harvest. Still, some farmers practice pre-cooling after the products are packed. Furthermore, there are some crops that can be successfully stored without pre-cooling. These include cabbage, cauliflower, green banana, guava, mango, papaya, pomegranate, and radish.

On the other hand, pre-cooling is an extremely valuable farm practice for various berries, stone fruits, peppers, cucumbers, tomatoes, melons, grapes, tangerines, okra, spinach, peas, and beans. Although certain crops can be stored without pre-cooling, farmers aiming towards the best results, practice pre-cooling as an integral part of their post-harvest management.

Crops that require pre-cooling.

The Choice of Pre-cooling Method

The choice of pre-cooling method will primarily depend on the requirements of each crop type, as well as on farmer's financial possibilities, required labor, available equipment, and materials. Once the crops are harvested, farmers have a big responsibility to do their best in order to protect the final yield and preserve crop quality. Pre-cooling is a simple but effective practice that reduces fruit respiration rate, and therefore, it's most definitely worth a try.

EVAPORATIVE COOLING CHAMBERS

Evaporative cooling chambers (ECCs), also known as “zero energy cool chambers” (ZECCs), are a type of evaporative cooler, which are simple and inexpensive ways to keep vegetables fresh without the use of electricity. Evaporation of water from a surface removes heat, creating a cooling effect, which can improve vegetable storage shelf life. ECCs are relatively large compared to the more common household clay pot cooler, and are therefore most suitable for farmers with large production quantities, farming groups, or farming cooperatives.

Suitability

ECCs provide the most benefits when they are used in low humidity climates (less than 40% relative humidity), the temperature is relatively hot (maximum daily temperature greater than 25 °C), water is available to add to the device between one and three times per day, and the device can be located in a shady and well-ventilated area.

Additionally, storage conditions must meet users’ needs for scale of storage needed and optimal conditions for different vegetables throughout the year. The cost of the ECC must be affordable and justified by the benefits be realized due to its improved storage.

Construction

The size of an ECC can be chosen to meet a range of user storage needs; however, the cost can vary significantly based on desired size and local cost of materials. Because ECCs can be constructed over a range of sizes, it is important to select an appropriate size according to the need, to avoid over-building and spending more money than is needed.

Evaporative cooling chambers (ECCs) can be made from locally available materials including bricks, sand, wood, dry grass, gunny/burlap sack, and twine. The space in between the two brick walls is filled with sand, which retains the water that is added. If the evaporative cooling chamber is not build in an area that is well shaded, a shed must be constructed to provide shade. Inside the ECC, food is placed in unsealed plastic containers, which keep the vegetables off the ECC’s floor and allows them to breathe and be exposed to the cool, humid air inside the device. Below is a detailed step-by-step process for building an ECC:

- Make a foundation of approximately 165 cm x 115 cm with bricks. The dimensions can vary according to the storage volume needed.
- Mount the double wall at approximately a height of 67.5 cm leaving a cavity of 7.5 cm wide between the walls.
- Wet the walls of the chamber with water.

- Fill the cavity between the two walls with wet sand. It is better to use the sand from a nearby river if possible.
- Make a frame from wood the same size as the foundation (165 cm 5) x 115 cm). Cover the wood frame with straw, dry grass, or burlap sack and secure with rope or twine.
- The cover goes over the evaporative cooling chamber to keep the cool air inside the chamber.
- Store your products in plastic crates or baskets in the chamber. Plastic crates containing products must be arranged on top of each other.
- Place a clear plastic sheet over the plastic baskets containing the fruits and vegetable ECC in operation to extend the shelf life of fruits and vegetables.
- If the evaporative cooling chamber is not build in an area that is well shaded, a shed must be constructed to provide shade.

Best Practices for Use

It is important that ECCs are correctly used to ensure maximum cooling performance benefit for the user. Improper use decreases the potential benefits and results in a lower cost-benefit ratio. Before starting to build an ECC, a location should be chosen that is close to water, exposed to wind/breeze, and if possible where there is shade to avoid the need of a cover. ECCs should be reinstalled every 3 years with new bricks. The cover of the ECC should be opened as infrequently as possible to keep the cool air in. The sand between the bricks must be kept wet, installing an irrigation system can make this process simpler. Additionally water should be sprinkled on the cover 1-3 times per day.

The vegetables that need storage should be carefully considered, since not all produce can be stored together because some release ethylene, which can accelerate ripening or reduce postharvest quality. Like any device for storing food, the ECCs should be kept clean. The surface of the interior cooling space should be sponged off regularly.

CURING

When roots and tubers are to be stored for long periods, curing is necessary to extend the shelf life. It is an effective operation to reduce the water loss during storage from hardy root and tuber vegetables such as potato, sweet potato, yam and other tropical vegetables where cuticle are poorly developed. They are relatively susceptible to mechanical wounding during harvesting and handling. These problems can be minimized by the process of 'curing' at intermediate to high temperature and high relative humidity (RH). During curing it develops periderm over cut, broken or skinned surfaces and helps in wound restoration. 'Curing is accomplished by holding the produce at high temperature and RH for several days, while harvesting wounds heals and new protective layers of cell forms around wound'.

Advantages:

- It helps in wound healing of harvest and handling injuries through skin hardening.
- Reduce water losses.
- Prevent infection from pathogen.

How Curing Happens?

- Curing is normally undertaken in the field, but in some case curing structures are employed.
- Produce can be cured in the field by piling them in a partially shaded area. Cut grass or straw can serve as insulating material, while, covering the pile with canvas, burlap, or woven grass matting. This covering will provide sufficient heat to reach high temperatures and high relative humidity. The stack can be left in this state for up to four days.
- Curing in potato starts with deposition of suberin in parenchymatous cell just below the damaged area of the tuber.
- Suberin (a waxy waterproof substance found in the cell walls of many plants, especially corky in nature) is a group of fatty acids which provides initial protection to the tuber against water loss and infection.
- Subsequently, below the suberized cells a meristematic layer of cells is formed which is the periderm, also called as cork cambium. This produces new cells which seal off the damaged area. But these processes are temperature and humidity dependent. Eg. Curing of potato takes place in 1 days at 21 °C; 2 days at 15 °C; 3 days at 10 °C; 5-8 days 5 °C.

FRUIT WAXING

Fruit waxing is the process of covering fruits (and, in some cases, vegetables) with artificial waxing material. Natural wax is removed first, usually by washing. Waxing materials may be either natural or petroleum-based. Potentially allergenic proteins (peanut, soy, dairy, wheat) may be combined with shellac.

The primary reasons for waxing are to prevent water loss (making up for the removal in washing of the natural waxes in fruits that have them, particularly citrus but also, for example, apples) and thus retard shrinkage and spoilage, and to improve appearance. Dyes may be added to further enhance appearance, and sometimes fungicides. Fruits were waxed to cause fermentation as early as the 12th or the 13th century; commercial producers began waxing citrus to extend shelf life in the 1920s and 1930s. Aesthetics (consumer preference for shiny fruit) has since become the main reason. In addition to fruit, some vegetables can usefully be waxed, such as cassava. A distinction may be made between storage wax, pack-out wax (for immediate sale), and high-shine wax (for optimum attractiveness).

Produce that is Often Waxed

A number of sources list the following as produce which may be waxed before shipping to stores: apples, avocados, bell and hot peppers, cantaloupes, cucumbers, eggplant, grapefruit, lemons, limes, mangoes, melons, nectarines, oranges, papayas, parsnips, passion fruit, peaches, pears, pineapple, plums, pumpkins, rutabaga, squash, sweet potatoes, tangarines, tomatoes, turnips and yucca.

Materials

The materials used to wax produce depend to some extent on regulations in the country of production and/or export. Both natural waxes (carnauba, shellac, or resin) and petroleum-based waxes (usually proprietary formulae) are used, and often more than one wax is combined to create the desired properties for the fruit or vegetable being treated. Wax may be applied in a volatile petroleum-based solvent but is now more commonly applied via a water-based emulsion. Blended paraffin waxes applied as an oil or paste are often used on vegetables.

POSTHARVEST HEAT TREATMENTS

Heat treatment substitutes a non-damaging physical treatment for chemical prevention. There are three methods in use to heat commodities; hot water, vapor heat and hot air. Hot water was originally used for fungal control, but has been extended to disinfestation of insects Vapor heat was developed specifically for insect control, and hot air has been used for both fungal and insect control and to study the response of commodities to high temperature. The last two methods (vapor heat and hot air) have subdivisions in that sometimes the air is relatively static, and sometimes air flow is quite high; additionally, hot air can have humidity control or not. All of these permutations may affect the response of the commodity to the heat treatment and affect the length of time of exposure needed to achieve a desired effect.

Heat Treatments

Hot Water Dips and Sprays

Hot water dips are effective for fungal pathogen control, because fungal spores and latent infections are either on the surface or in the first few cell layers under the peel of the fruit or vegetable. Postharvest dips to control decay are often applied for only a few minutes, at temperatures higher than heat treatments designed to kill insect pests located at the interior of a commodity, because only the surface of the commodity requires heating.

Many fruits and vegetables tolerate exposure to water temperatures of 50–60 °C for up to 10 min, but shorter exposure at these temperatures can control many postharvest plant pathogens. In contrast, hot water dips for fruit require 90 min exposure to 46 °C.

Fungicide effectiveness can be enhanced by applying the fungicide in a hot water bath, thus allowing more effective fungal control with a reduction in chemicals. This has been particularly effective on citrus with the fungicides thiabendazole and imazalil. In addition, generally recognized as safe (GRAS) compounds have been applied in hot water to improve the efficiency of their antifungal action. Heated solutions (45 °C) of sulfur dioxide, ethanol or sodium carbonate have been used to control green mold (Penicillium digitatum) on citrus fruits. These compounds were as effective as an imazalil dip at 25 °C in controlling artificial inoculations of the fungus.

A recent extension of the hot water treatment has been the development of a hot water spray machine. This is a technique designed to be part of a sorting line, whereby the commodity is moved by means of brush rollers through a pressurized spray of hot water. By varying the speed of the brushes and the number of nozzles spraying the water, the commodity can be exposed to high temperatures for 10-60 s. The water is recycled, but because of the temperatures used (50–70 °C), organisms which are washed off the product into the water do not survive.

This machine is in use both to clean and to reduce pathogen presence on a number of fruits and vegetables, such as mangos and peppers. Packers receive a premium for using this machine on these commodities as well as on corn and melons.

Hot water dips have been used for disinfesting insects as well. Since hot water is a more efficient heat transfer medium than hot air, when properly circulated through a load of fruit a uniform temperature profile is established in the bath. For disinfestation a longer treatment is necessary than for fungal control because the total fruit and not just the surface has to be brought to the proper temperature. Procedures have been developed to disinfest a number of subtropical and tropical fruits from various species of fruit fly. The times of immersion can be 1 h or more and temperatures are below 50 °C, in contrast to many antifungal treatments which are for minutes at temperatures above 50 °C.

Vapor Heat

Vapor heat is a method of heating fruit with air saturated with water vapor at temperatures of 40–50 °C to kill insect eggs and larvae as a quarantine treatment before fresh market shipment. Heat transfer is by condensation of water vapor on the cooler fruit surface. This procedure was first used to kill Mediterranean (Ceratitis capitata Wiedemann) and Mexican (Anastrepha ludens Loew) fruit fly in a chamber without forced air. However, once ethylene dibromide and methyl bromide came into use as inexpensive chemical fumigants, vapor heat was abandoned. With the ban on use of ethylene dibromide in 1984, and the imminent removal of methyl bromide from use in 2010, vapor heat has again come into use. However, in modern facilities the vapor heat includes forced air which circulates through the pallets and heats the commodity more quickly than vapor heat without forced air. Commercial facilities operate in many countries, mainly for use on subtropical fruits, particularly mango and papaya. In addition, studies have been conducted for using vapor or moist forced air to disinfest many fruits and vegetables from various insect pest.

The treatment consists of a period of warming (approach time) which can be faster or slower depending on a commodity's sensitivity to high temperatures. Then there is a holding period when the interior temperature of the produce reaches the desired temperature for the length of time required to kill the insect. The last part is the cooling down period which can be air cooling (slow) or hydrocooling (fast). Thus, there are a number of components of the treatment which can be manipulated to find the best combination for elimination of the insect pest without damaging the commodity.

Hot Air

Hot air can be applied by placing fruit or vegetables in a heated chamber with a ventilating fan, or by applying forced hot air where the speed of air circulation is precisely controlled. Hot air, whether forced or not, heats more slowly than hot water immersion or forced vapor heat, although forced hot air will heat produce faster than a regular heating chamber. The hot air chamber has been utilized to study physiological changes in fruits and vegetables in response to heat.

Forced hot air, however, has been used to develop quarantine procedures. One reason is that the high humidity in vapor heat can sometimes damage the fruit being treated, while the slower heating time and lower humidity of forced hot air can cause less damage. A high temperature forced air quarantine treatment to kill Mediterranean fruit fly, melon fly and oriental fruit fly on papayas has been developed. This procedure may require rapid cooling after the heat treatment to prevent fruit injury, as may the forced hot air treatment for citrus. Recently, the heat treatment for papaya has been modified so that hydrocooling is not necessary. This method is under investigation for other commodities such as persimmons.

Exposure to high temperature forced or static air can also decrease fungal infections. Heating without forced air can reduce decay caused by Botrytis cinerea and Penicillium expansum in apple fruit and Botrytis cinerea in tomatoes. The treatments used in these cases are long term heating, from 12 to 96 h at temperatures ranging from 38 to 46 °C, and are unlikely to become a commercially attractive treatment. However, the potential to have in hot air treatment a means of beneficially affecting commodity physiology and at the same time preventing both insect and fungal invasion justifies further development of these treatments.

Commodity Responses

Flowers

Not being eaten, flowers are not subject to the same restrictions concerning chemical residues as other commodities. On the other hand, with increasing international trade flowers are subjected to the same international restrictions on the import of insects as other commodities. Current postharvest approaches are hand removal, insecticidal dips or fumigation, biological control agents and temperature treatments. A recent review discusses the advantages and disadvantages of these treatments.

Both hot water dips and forced vapor heat have been used to kill insects on flowers. Hot water treatments for flowers or cut foliage can either be for hours at temperatures of 42 °C or below, or for minutes at temperatures of 46 °C or above. Aphids in red ginger were destroyed after 5 min in a 47 °C water bath without plant injury. Hara et al. achieved quarantine security against magnolia white scale on bird of paradise flowers by immersion in 49 °C water for 10 min. Flower vase life was reduced by one or two days, depending on the stage of flower opening at the time of treatment. Hot water has been investigated against a number of insects in flowers, foliage and roots and all were controlled by 49 °C for 5–12 min. Some flowers showed greater susceptibility to heat injury at different seasons, and preconditioning in 39 °C hot air for 2–4 h before the hot water treatment eliminated this seasonal phytotoxicity.

Hot water dips can also be used to control fungal pathogens. Hot water treatments at temperatures around 40 °C for several hours have been used to treat bulbs, seeds and other planting material against diseases. Dipping roses for 30 s at 50 °C was effective in preventing Botrytis cinerea development. Botrytis can also cause stem blight on propagated cuttings, and intervals of hot forced air under a greenhouse bench reduced the incidence of the disease.

Forced vapor heat can also be used successfully to control insects on tropical cut flowers and foliage, although in many cases the vase life is reduced by the length of time required for full insect mortality. The same strictures apply for flowers as for other commodities; to find a time–temperature treatment which will destroy the insect or fungal pest without harming the plant tissue. Unlike fruits and vegetables, there have been no reports of heat treatments that improve vase life.

Fruit Ripening

Ripening of most climacteric fruit is characterized by softening of flesh, an increase in the sugar:acid ratio, enhanced color development, and increases in respiratory activity and ethylene production. Exposing fruit to high temperatures attenuates some of these processes while enhancing others. This anomalous situation results in heated fruit being more advanced in some ripening characteristics than non-heated fruit while maintaining their quality longer during shelf life at 20 °C.

The inhibition of ripening by heat may be mediated by its effect on the ripening hormone, ethylene. Hot air treatment of 35–40 °C inhibits ethylene synthesis within hours in both apples and tomatoes. Elevated temperatures of 35–38 °C can cause endogenous ACC to accumulate in apple and tomato tissue concomitantly with the decrease in ethylene, though raising the temperature higher or holding the fruit longer at the raised temperature will cause the disappearance of ACC as well. A rapid loss of ACC oxidase activity occurs in many fruit exposed for a few hours to hot water immersion at 42–46 °C, due primarily to decrease in ACC oxidase mRNA and cessation of enzyme synthesis. ACC synthase is also heat labile, but most studies indicate that it is less heat sensitive than ACC oxidase. The inhibition of ethylene formation is reversed when the fruits are removed from heat, and often the level of ethylene rises to higher levels than in non-heated fruits. This recovery requires protein synthesis, and studies showed that both mRNA and protein of ACC oxidase accumulate during recovery from a 38 °C hot air treatment.

During the heating period, not only is endogenous ethylene production inhibited, but fruits will not respond to exogenous ethylene this indicates either a loss or inactivation of ethylene receptors, or the inability to transfer the signal to the subsequent series of events leading to ripening. No information is available on the response of ethylene receptors to heat, but it has been shown that the expression of tomato ripening genes is inhibited by high temperature. Specific mRNAs associated with ripening processes were found to disappear during a 38 °C hot air treatment of tomatoes and reappear during recovery from heat. These included ACC oxidase, polygalacturonase and lycopene synthase.

Fruits subjected to extended hot air treatments of 38 or 40 °C often soften more slowly than non-heated fruits, although disinfestation procedures for mangos and papaya of hot forced air for 4 h at 50°C led to faster softening after the treatment. A number of researchers have described the effect of continuous storage in hot air on fruit firmness. Plums, pears, avocados, and tomatoes softened more slowly when held continuously at temperatures between 30 and 40 °C than at 20 °C. The rate of softening increased when heated fruit were returned to 20 °C, but it was still less than that of non-heated fruit. Even after 6 months of storage at 0 °C and subsequent shelf life for 7 days at 20 °C, apples that had been held at 38 °C for 3 or 4 days pre-storage were 10 N firmer than non-heated fruit. The texture of hot air treated apples after storage was different quantitatively and qualitatively from non-heated fruit. Conway et al. using compression tests found the heated apples to be tougher, while Lurie and Nussinovich, using Instron compression and shearing measurements, found heated apples to be crisper than non-heated.

Cell wall studies of apple fruit found less soluble pectin and more insoluble pectin after exposure to 38 °C air for 4 days than in fruit that had not been heated, an indication of inhibition of uronic acid degradation.

In addition, in these heated apples less calcium was present in the water soluble pectin and more was bound to the cell wall. It was thought that this was the result of the activity of pectin esterase creating more sites for calcium binding, but a study of heated and non-heated fruits showed a similar degree of esterification in both. During the heating period arabinose and galactose content decreased with no accompanying decrease in uronic acids. It is possible that loss of neutral sugar side chains during the heat treatment may lead to closer packing of the pectin strands and in turn hinder enzymic cleavage during and after storage, resulting in firmer fruit.

The decrease in the rate of softening may be due to inhibition of the synthesis of cell wall hydrolytic enzymes such as polygalacturonase and a- and b-galactosidase. In tomato mRNA for polygalacturonase was absent in fruit during a heat treatment of 1–3 days at 38 °C and appeared after the fruit was removed from heat. Depending on the length of treatment, heated tomato fruits may recover and soften to the same extent as non-heated fruits, or remain firmer than non-heated fruits. In the former study tomatoes were held for 3 days at 38 °C and in the latter for 4 days at 40 °C.

Flavor characteristics of fruits can be affected by a heat treatment. Titratable acidity declines in apples held for 3 or 4 days at 38 °C while soluble solids concentration is not affected by the treatment. The same was found after hot forced air treatment of nectarines at 41–46 °C for 1–2 days for insect disinfestation, and hot water immersion for 15 min at 35, 45 or 55 °C of strawberries for decay control. In tomatoes hot air heated at 38 °C for 2–3 days and grapefruit held in forced hot air of 43.5 °C for 4.5 h, neither titratable acidity nor soluble solids content was affected by heat. However, the same fruits in other studies showed reduction in titratable acidity. The disparate results may be due to cultivar differences or differences in the heat treatment.

In some commodities, sugar content is favorably affected by heat treatment. For example, 3 h of 45 °C water before cool storage of muskmelons prevented the loss in sucrose which occurred in non-heated fruit during storage. Squash sucrose content can also be raised by holding them at 30 °C in air before storage. These heat treated squash were perceived as sweeter by a taste panel. Heated tomatoes were not distinguished from non-heated by a taste panel, but heated Golden Delicious apples (4 days at 38 °C) were perceived as crisper, sweeter and overall more acceptable than non-heated fruit. In the latter fruit the sweetness was due more to decrease in acidity than increase in sugar content.

Volatiles production can also be affected by a heat treatment of hot water immersion at 42 °C for 60 min or hot air for 2 days at 38 °C. Volatiles production in apples is enhanced during a 38 °C hot air treatment, is inhibited immediately following the treatment, and recovers afterwards. The profile of the volatiless is also changed, with some being enhanced more than others by the heat. In tomatoes as well the highest volatiles levels in ripe fruit were from fruit heated at the mature green stage and then stored at 13 °C before ripening.

Heat treatment leads to an accelerated rate of degreening in apples. Chlorophyll content in apple peel, plantain peel and tomato pericarp decreased during a hot air treatment of 35–40 °C. Hot water immersion at 45 °C for 30–60 min can also lead to yellowing of cucumbers, as does forced vapor heat for 30 min at 45 °C for zucchini (zucchini are known as courgettes in some countries). Color changes in papaya skin or flesh were not affected by hot water immersion at 42 °C for 30 min followed by 49 °C for 90 min and the same hot air treatment which stimulated degreening of plantains failed to degreen bananas. Hot water dips at 43–55 °C for up to 10 min delayed yellowing of broccoli. The difference in responses of different commodities may be an indication of whether new enzymes must be synthesized to effect the color changes or not. In the case of apples chlorophyll degradation reveals the yellow of the underlying carotenoids already present, while other fruits may require synthesis of carotenoids. For example, it has been found that hot air at 38°C or higher inhibits lycopene synthesis in tomatoes. The inhibition of lycopene is due to the inhibition of transcription of mRNA for lycopene synthase, a key enzyme in the pathway, and this recovers after removal from heat. In bananas the inhibition of degreening during the heat treatment appears to be due to the absence of the chlorophyll oxidase enzyme resulting in the retention of chlorophyll in the peel. It is not known if this inhibition is at the level of gene expression.

Respiration rate is enhanced initially for the first day or two by high temperatures of 35–40 °C, but at longer times at high temperatures the rate decreases. With increasing time at 35 °C a greater proportion of the respiration is from the cyanide insensitive pathway. When fruits return to ambient temperature often the respiration is lower than non-heated fruits. A heat treatment, depending on temperature and length of exposure, can decrease or increase the climacteric respiration peak as well as advancing or delaying it after treatment. The response of a particular fruit or vegetable will result from a combination of factors: preharvest environmental conditions, the physiological age of the commodity, the time and temperature of exposure, whether the commodity is removed from heat to storage or to ripening temperature, whether the heat treatment causes damage.

Thermotolerance

The mechanism by which a heat treatment causes changes in fruit ripening, such as inhibition of ethylene synthesis, and cell wall degrading enzymes, may be tied to changes in gene expression and protein synthesis. During a high temperature treatment the mRNA of fruit ripening genes disappear and those of heat shock proteins (HSP) accumulate. An immediate response of high temperature, generally temperatures above 35 °C, is disassociation of polyribosomes and then a reassociation of some ribosomes into polyribosomes which preferentially translate the mRNA of HS. This response both downregulates normal protein synthesis even without degradation of the mRNAs and upregulates HSP synthesis. The synthesis of HSP is part of the response of all organisms to a heat stress, from man to bacteria. Studies with many organisms have demonstrated that exposure to elevated, sublethal temperatures induces thermotolerance, which protects them from a second exposure to a normally lethal temperature. Development of thermotolerance has been associated with synthesis of HSP, and loss of thermotolerance with the disappearance of HSP. In addition, the development of thermotolerance is dependent on protein synthesis; pepper discs treated with protein synthesis inhibitors before temperature treatments did not develop thermotolerance.

Therefore, heat exposure severity will moderate the thermotolerance response. Development of thermotolerance is dependent on exposure temperature. The exposure temperature must be enough to initiate the synthesis of HSP, yet not too hot so that transcription and translation of HSP are inhibited. Temperatures of 35–40 °C have been found to be effective, depending upon the commodity. At 42 °C or higher, HSP synthesis is attenuated and commodities are more likely to suffer heat damage.

The propensity of a medium heat stress to protect against a higher heat stress has been used to develop treatments to prevent commodity damage while killing fungal pathogens or insect pests. A two-stage hot water treatment, 30 min at 42 °C followed by immersion in 49 °C water, was developed for papaya disinfestation. Holding the fruit at 38–42 °C for 1 h reduces damage after hot water immersion at 49 °C for 70 min. In a similar way, conditioning in 37 or 39 °C air before a 46° or 47 °C hot water disinfestation reduces treatment damage on avocados and mangos. Similar benefits of prior temperature conditioning have been found for avocados and cucumbers.

Tolerance to Chilling Injury

The correlation between HSP and thermotolerance has been established in many organisms, but only recently has it been found that a heat stress can condition plants to low temperature. Salveit and coworkers found that prior high temperature exposure of a number of hours at 38–42 °C hot air affected chilling sensitivity of tomato discs, mung bean hypocotyls and cucumber cotyledons and seeds. When a heat treatment of 2–3 days in 38 °C air was applied to tomato fruit their sensitivity to low temperature was reduced and they could be stored for up to a month at 2 °C without developing chilling injury. This resistance to low temperature injury was found to be contingent on the presence of HSP. In a study with avocado discs, maximal HSP production was found after 4 h at 38 °C and heating provided a significant degree of protection from chilling injury. This response has been found in numerous other commodies including avocado, citru, cucumber, mango, pepper, persimmons and zucchini. However, the response may in some cases be cultivar specific. For example, Whitaker found no benefit in heating 'Rutgers' tomato fruit.

Other examples of the effectiveness of a heat exposure in reducing chilling sensitivity are the use of a 12–18 h 38 °C hot air treatment together with cold quarantine to disinfect avocado from fruit fly without having the fruit develop chilling injury. When the heat treatment was given as a fungicidal hot water dip, rots were also controlled.

The reduction of sensitivity to chilling injury of fruits may not be due solely to the presence of HSP. Chilling injury has long been thought to begin with membrane damage, and a heat treatment of 35–40 °C may cause membrane alterations. High temperature increases membrane leakage, but after removal from heat stress the tissue recovers and leakage returns to levels found in tissue held at 20 °C. Using membrane leakage as a measure of chilling injury Saltveit found that conditioning tomato fruit discs at 37 °C for 4 h reduced leakage when discs were stored at chilling temperature. An examination of the lipid composition of apple plasma membrane showed that after a 4 day heating in 38 °C air and 0 °C cold storage for 4 months there were more phospholipids and greater fatty acid unsaturation in heated than in non-heated fruits. Whitaker et al. looking at total apple lipids also found greater fatty acid unsaturation though not greater phospholipid content in fruit heated by the same method. This would indicate more fluid membranes in fruit after exposure to conditioning temperatures, and correspond with lower indiscriminate leakage from conditioned fruit and vegetable tissue. These changes in lipid composition were found also after storage at 2 °C in tomatoes treated prestorage for 2 days at 38 °C with hot air, or immersed in 46 and 48 °C water for 2–3 min, an indication that even short exposure to heat can instigate processes leading to tissue adaptation to low temperature.

Apples are normally thought to be fruits which are insensitive to low temperature, but superficial scald is a physiological storage disorder which is a form of chilling injury. It is an oxidative process causing peel browning, and has been correlated with the oxidation of a-farnesene, a component of the apple wax. A heat treatment of 3–4 days at 38 °C of apples before storage, controls this disorder during the first months of 0 °C storage by inhibiting the accumulation of a-farnesene and consequently decreasing the oxidation products. The inhibitory effect allows for 3–4 months of air storage without scald developing. Part of the overall reduction in a-farnesene may be due as well to a thinner wax layer and changes in structure of the wax surface after the heat treatment.

Heat Damage

This is one reason why there are such a multitude of treatments, to find a time–temperature regime which will produce the desired effect (disinfestation, fungal control) without damaging the commodity. Damage can be both external and internal. External is generally peel browning, pitting, or yellowing of green vegetables such as zucchini or cucumber. Tissue damage caused by heat will also result in increased decay development. Internal damage can evidence itself in mango and papaya as poor color development, abnormal softening, the lack of starch breakdown and the development of internal cavities. In addition, the fruit can soften quickly or show abnormal softening where some areas of the flesh remain hard while others soften. Internal damage on other fruits can include flesh darkening on lychee and nectarines. If the produce is stored after the heat treatment at low temperature the heat damage can be confused with chilling injury which has similar symptoms.

SANITATION

Whole fresh fruits and vegetables are exempt from legally required implementation of GMPs, but the focus on sanitation practices is good for reducing food safety risks in fresh produce operations. Adopting good postharvest practices will not only reduce food safety risks, but also contribute to maintaining produce quality and reducing postharvest decay. To understand what risks may be present, an assessment of the packing and produce handling area should be completed. Any surface that the produce touches may serve as a source for contamination. Such surfaces include equipment, belts, rollers, brushes, tables, bins, sinks, tools, and even the hands of workers. Surfaces that contact produce must be able to be easily cleaned and preferably, sanitized.

Areas within the packing and handling area can be broken into zones to help determine the likelihood of direct contact with the produce you are handling:

- Zone 1: Direct food contact surfaces such as conveyors, belts, brushes, rollers, sorting tables, racks, utensils, harvest/storage bins, and worker hands. This zone is the biggest concern because it has direct contact with the produce and if contaminated, could result in contamination of the entire crop.
- Zone 2: Non-food contact surfaces that are in close proximity to the product, such as internal and external parts of washing or processing equipment such as sidewalls, housing, framework, or spray nozzles.
- Zone 3: Areas inside of the packing area such as trash cans, cull piles, floors, drains, restrooms, forklifts, phones, and catwalks or storage areas above packing areas.
- Zone 4: Areas outside of or adjacent to the packing area such as loading docks, warehouses, manure or compost piles, and livestock operations.

To prioritize resources and address the most likely risks of contamination, begin addressing risks in Zone 1 since these are direct food contact areas, then proceed to Zones 2, 3, and 4. To help identify the zones in your packing area, diagram the flow of produce through the packinghouse and add zone numbers to each area.

The cleaning and sanitizing process includes four steps:

- Step 1: The surface should be rinsed so any obvious dirt and debris are removed.
- Step 2: Apply an appropriate detergent and scrub the surface.
- Step 3: Rinse the surface with water that is the microbial equivalent of drinking water (potable).
- Step 4: Apply an appropriate sanitizer. If the sanitizer requires a final rinse, this will require an extra step. Let the surface air dry.

Writing Standard Operating Procedures (SOPs) will help outline what areas need to be cleaned, how often, what detergents and sanitizers to use, how to clean and/or sanitize the surface, and who is responsible for completing the task.

Different challenges may exist depending on the type of structure where fruits and vegetables are packed and handled. Whether packing in an open tent or enclosed packinghouse, it is important to assess risks and minimize them. Pests are one obvious risk of contamination to fresh produce. Birds, rodents, and insects should be deterred or removed from all packing and produce handling areas using the best available methods such as setting traps, installing fencing and netting, closing doors, and repairing any holes in windows. It is critical that packing and sorting areas be cleaned of debris after each day to prevent attracting pests, such as rodents. A pest control program, including SOPs for monitoring and correcting any pest problems, is a valuable part to any food safety program.

In addition to keeping buildings, equipment, and surfaces clean, worker education and training will reduce the risk of postharvest contamination. Training should include any glove/apron policies, proper handwashing and bathroom use, cleaning and sanitizing tasks, and eating and drinking policies. Trained workers are a valuable resource to identify factors that increase fruit damage. Damaged or cut fruit can harbor human pathogens and decay organisms as well as provide a growth medium for pathogens on food contact surfaces. Workers may also be able to identify other food safety risks while they are working, including animal fecal contamination or other visible contamination of the crop being handled.

References

- Postharvest-processing, agricultural-and-biological-sciences: sciencedirect.com, Retrieved 2 January, 2020
- Proper-postharvest-cooling-and-handling-methods: content.ces.ncsu.edu, Retrieved 19 June, 2020
- Drying-as-a-key-process-in-post-harvest-technology, post: agrivi.com, Retrieved 1 March, 2020

- John M. Krochta, Elizabeth A. Baldwin and Myrna O. Nisperos-Carriedo, Eds., Edible Coatings and Films to Improve Food Quality, Lancaster, Pennsylvania: Technomic, 1994, ISBN 9781566761130
- Sanitation-and-postharvest-handling, decision-trees, educational-materials: gaps.cornell.edu, Retrieved 21 May, 2020

3

Postharvest Handling of Fruits and Vegetables

It is important to ensure the fruits and vegetables are handled with utmost care in the postharvest period. This chapter has been carefully written to provide an easy understanding of the various facets of postharvest handling of fruits and vegetables. Some of the fruits and vegetables focused in this chapter are apples, mangoes, tomatoes and beans among others.

PROCESSING AND PRESERVATION OF FRESH-CUT FRUITS AND VEGETABLES

Fresh-cut fruit and vegetable products (FFVP) are defined as fruits or vegetables that have been trimmed, peeled or cut into a 100% usable product which has been packaged to offer consumers high nutrition and flavour while still maintaining its freshness. The importance of fresh-cut produce lies in its major characteristics of freshness, convenience, nutrient retention and sensory quality while providing extended shelf life. Fresh-cut fruits and vegetables (FFV) are products partially prepared and which require no additional preparation for their use. This makes it unavoidable that their overall quality diminishes during processing and storage. It is made more so, as the operations involved in preparing fresh-cut products damage the integrity of the cells, promotes contact between enzymes and substrates, increases the entry of microorganisms and creates stress conditions on the fresh-cut produce.

FFVP are also referred to as products prepared with slight peeling, cutting, shredding, trimming and sanitizing operations and which have been packed under semipermeable films and stored under refrigerated temperature. Fresh-cut products are also reported to contain similar nutrients and ingredients as whole products with the added advantage of short time preparation and the low prices at which they are been sold. Fresh-cut products constitute a major rapidly growing food segment which is of interest to food processors and consumers. The fresh-cut industry is expanding more rapidly than other sectors of the fruit and vegetable market due to its supply of both the food service industry, retail outlet as well as its expanding production and access to new markets across the globe. The growth rate of the sector is reported to be in the region of billions of dollars in recent years with USA as the main producer and consumer while the UK and France follows after. FFVP presently sold in markets across the globe includes: lettuce (cleaned, chopped, shredded), spinach/leafy greens (washed and trimmed), broccoli and cauliflower (florets), cabbage (shredded), carrots (baby, sticks, shredded), celery (sticks), onions (whole peeled, sliced, diced), potatoes and other roots (peeled, sliced), mushrooms (sliced), jicama/zucchini/cucumber (sliced, diced), garlic (fresh peeled, sliced) as well as tomato and pepper (sliced).

Despite the fact that food processing methods extend the shelf life of fruit and vegetable products, processing of fresh-cut produce however reduces the shelf life of the commodity, rendering the product highly perishable as a result. This biological change may lead to flavour loss, cut-surface discolouration, decay, rapid softening, increased rate of vitamin loss, shrinkage as well as shorter shelf life of the fresh-cut produce. Interactions between intracellular and intercellular enzymes with substrates as well as increased water activity may also lead to flavour and textural changes upon processing. A major effect of fresh-cut processing is stress on vegetable tissues with the resultant phytochemical accumulation and loss induced through reduced activity in key enzymes of secondary metabolic pathways. Fresh-cut processing also results in cell breakdown as well as the release of intracellular products such as oxidizing enzymes thereby quickening product decay.

Several factors are reported to affect the overall quality of fresh-cut produce. Among many of such factors is appearance. Appearance in combination with size, shape, form, colour as well as the absence of defects are factors which greatly affects the purchase of fresh-cut produce by consumers. All of these factors can also be influenced by several pre-harvest factors. Available nutrients inherent in fruits and made available upon consumption includes antioxidant vitamins beta-carotene (pro-vitamin A), α-tocopherol (vitamin E) and ascorbic acid (vitamin C). Research has also shown that regular consumption of fruits and vegetables reduces risk of cancers, cardiovascular diseases and several inflammations. This apart from regular body exercise and genetics has made fruit and vegetable consumption one of the main factors that contributes to a healthy lifestyle. With studies showing the nutritional benefits of fruits and vegetables, consumption of FFVP therefore promotes health through increase in the supply of antioxidant and other phytochemical nutrients to the body.

Processing of FFV

FFV have been known to have a shorter shelf-life compared to intact fruit and vegetable products due mainly to processing. Several processes involved in the production of FFVP have been known to alter greatly the shelf stability of the cut-produce. There are traditional processing procedures involved in obtaining fresh-cut products and this procedure usually requires an order of unit operations such as peeling, trimming, shredding, dicing, cutting, washing/disinfecting, drying and packaging. Shelf life extension of the cut produce is therefore dependent on a combination of these unit operations as well as proper temperature management during storage, use of antibrowning agents, proper packing conditions as well as good manufacturing and handling practices. The unit operations required in the handling and processing of FFVP is shown in figure.

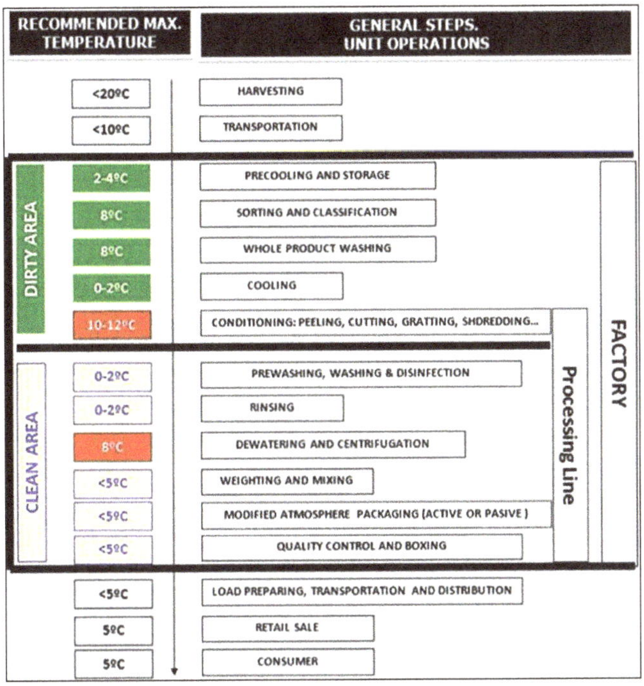

Unit operations and maximum recommended temperatures for each step in the processing of fresh-cut fruit and vegetable produce.

Cutting

An essential aspect of processing of fresh-cut produce is cutting. Cutting helps divide whole harvested fruit and vegetable products into minute fractions before packaging. The effect of cutting however on the products is the wounding stress which the cut tissues are allowed to suffer thus accelerating the rate of spoilage and deterioration of the cut produce. Cutting has been attributed to be the main factor responsible for the deterioration of FFV thereby enabling the product to experience a more rapid rate of deterioration than whole products. Cutting increases respiration rate, induces deteriorative changes associated with plant tissue senescence and thus the consequential decrease in shelf life when compared to the unprocessed produce. Cutting shape as well as the sharpness of the cutting blade has been attributed as some factors that affect the quality attributes of fresh-cut products. The works of Portela and Cantwell showed that melon cylinders cut with a blunt blade demonstrated higher ethanol concentrations, off-odour scores, electrolyte leakage, and increased potential for ethylene secretion when compared to products processed with a sharp blade. It was also reported that use of sharp cutting implements reduces wound response, lignin accumulation, white blush, softening and microbial growth in fresh-cut carrots. Cutting-induced injury has been implicated as affecting the immediate visual quality of fresh-cut products and has also been known to have longer-term effects on metabolism with the concomitant quality changes that are detected at a later time. The actual cutting process results in great tissue disruption as formerly sequestered enzymes and substrates mix are found to mix, hydrolytic enzymes released, while signalling-induced wounding responses may be initiated.

During the process of cutting, phenolic metabolism takes place: breakage of the plasma membrane with the resultant effect of inducing oxidative enzymatic reactions thus triggering browning of tissues and oxidation of polyphenols; and production of injury signals which induces the secretion of more secondary metabolites including phenolic antioxidants to heal the wound damage. It has been reported that the content of phenolic acids increases in fresh-cut products. This fact can be attributed to the cutting of fresh fruits and vegetables with knives thereby inducing the activity of polyphenol oxidase (PPO) in the cut fresh fruit and vegetables. FFV are thus easily susceptible to browning reaction as a result. Accumulation of phytochemicals can also be as a result of altered O_2 and CO_2 levels during packaging as well as the use of preservatives such as ascorbic and citric acid.

Wounding as a result of cutting has been attributed as one of the basic source of stress experienced by fresh-cut produce. Some factors can however affect the wound response of the fresh-cut produce and these factors include stage of maturity, cultivar, storage, processing temperature, cutting method, water vapour pressure as well as O_2 and CO_2 levels. Wounding stress as a result of cutting of fruit and vegetables has been shown to increase the antioxidant activity as well as the polyphenolic content in fresh-cut produce such as carrots, celery, lettuce, broccoli, mushroom, onions, and mangoes. Consequences of wounding includes increase in respiration rate; production of ethylene; oxidative browning; water loss; and degradation of membrane lipids. This therefore increases the susceptibility of FFV to increased perishability than their source commodity.

Sanitation and Hygiene in Processing Facilities

During the production process, cut fruits are exposed to environmental microbes in the processing facility. Reducing the level and rate of contamination will be dependent on the use of the appropriate disinfectants and sanitizers. One of such disinfectants of high use in the FFV industry is chlorine. The use of chlorine as a disinfectant is however of great concern and is presently prohibited in some European countries due to issues of public health. Chlorine is generally used in the food industry due mostly to its low price and its wide application of antimicrobial effectiveness. However, under certain conditions, chlorine has been shown to be weak in reducing microbial loads as it can easily be inactivated by organic matter and its action is highly pH reliant. Chlorine has also been shown to produce unhealthy by-products which are carcinogenic and mutagenic such as chloroform, trihalomethanes, chloramines and haloacetic acids, when reacting with organic molecules. Chlorine is also corrosive with its use banned in some European countries such as Belgium, Denmark, Germany and The Netherlands. Presently, alternative chemical compounds, biological methods and physical technologies which are more environmentally friendly and possess less risk to the health of workers and consumers have been developed to replace the use of chlorine.

Concentrations of 50–200 ppm and exposure time of 5 min of chlorine is commonly applied as hypochlorous acid and hypochlorite and as disinfectant in the FFV industry in order to enhance microbial safety of the produce. The exposure time of 5 min (depending on the microorganism) has been shown as the maximum exposure time required as longer times of > 5–30 min did not result in increased removal of the pathogenic organisms.

In the handling and processing of FFV, common practices are undertaken and needs to be taken note of. These practices consist of protection from damage as a result of poor handling and poor functioning of machinery, foreign body contamination, improper washing, drying and unhygienic practices by personnel. Hence worker sanitation which is most often neglected in the fresh cut industry in collaboration with good manufacturing practices must be enforced by food processors. In accompanying this process, training of food handlers in food hygiene techniques must be undertaken.

Presently, new and alternative technologies for safety, improved quality and extended shelf life of processed fresh-cut products have been developed. Such technologies include: ozone (O_3), a strong oxidizing agent in destroying microorganisms which has also been suggested as an alternative to sanitizers due to its effectiveness at low concentrations, short contact times and in the breakdown of nontoxic products; chlorine dioxide (ClO_2), which is known for its efficacy against pathogenic spores, bacteria and viruses; organic acids and calcium (Ca) salts applied for maintenance of cell wall structure and firmness (Ca), inhibition of enzymatic and non-enzymatic browning as well as in the prevention of microbial growth at heights that did not affect flavour of the fresh-cut products with their efficacy against microbes higher for bacteria than molds; electrolyzed water employed due to its strong bactericidal effect against pathogens and spoilage microbes.

Inhibition of Browning in FFV

It has been shown that fresh-cut process increases the metabolic activity mainly as a result of the enzymes polyphenol oxidase (PPO) causing discoloration and peroxidase (POD) causing enzymatic browning as well as de-compartmentalization of enzymes and substrates in tissues causing changes in flesh colour. PPO can induce the browning occurrence by catalyzing the oxidation of phenol to o-quinones which are polymerized to produce brown pigments. Postharvest techniques maintaining the quality of fresh-cut fruit have been investigated by several researchers including physical and chemical treatments. Many anti-browning agents or mixtures have been investigated like: calcium ascorbate with citric acid and N-acetyl-L-cysteine, citric acid, ascorbic acid with citric acid and calcium chloride, 4-hexylresorcinol with potassium sorbate and D-isoascorbic acid and modelling of the effects anti-browning agents on colour change in fresh-cut.

The effects of different anti-browning compounds (ascorbic acid, citric acid, L-cysteine and glutathione) at four concentrations of 0% as control, 0.5, 1.5 and 2.5% on L*-value, hue angle, brown scores and brown pigments of fresh-cut mangoes were investigated by Techavuthiporn and Boonyaritthonghai and they observed similar changing tendency of L*-value and hue angle decreasing in time during storage at 10 °C, while the brown scores and amount of brown pigment increased. They also observed that treatment with L-cysteine or glutathione was effective in suppressing tissue metabolism, PPO and POD activities, while citric acid significantly inhibited the growth of microorganisms.

Table: The estimated parameters k and the standard error of estimates (S.E.) for dipped fresh-cut mangoes at different concentration and anti-browning agents.

Treatments	K_L ± S.E. (day^{-1}) a	K_H ± S.E. (day^{-1}) a	K_{BS} ± S.E. (day^{-1}) a	K_{BP} ± S.E. (day^{-1}) a
Control	0.0209 ± 0.0012	0.0158 ± 0.0008	0.2604 ± 0.0474	0.2616 ± 0.0187
Ascorbic acid				
0.5%	0.0132 ± 0.0012	0.0132 ± 0.0002	0.3313 ± 0.04567	0.1566 ± 0.0231
1.5%	0.0088 ± 0.0007	0.0090 ± 0.0004	0.2146 ± 0.0777	0.1272 ± 0.0062
2.5%	0.0066 ± 0.0010	0.0079 ± 0.0003	0.2231 ± 0.0572	0.1186 ± 0.0129
Citric acid				
0.5%	0.0213 ± 0.0023	0.0120 ± 0.0014	0.1859 ± 0.0466	0.1981 ± 0.0212
1.5%	0.0210 ± 0.0030	0.0098 ± 0.0012	0.1642 ± 0.0971	0.1559 ± 0.0177
2.5%	0.0164 ± 0.0023	0.0097 ± 0.0007	0.1443 ± 0.1053	0.1386 ± 0.0062
L-Cysteine				
0.5%	0.0068 ± 0.0019	0.0095 ± 0.0011	0.1776 ± 0.0521	0.1427 ± 0.0263
1.5%	0.0053 ± 0.0008	0.0077 ± 0.0011	0.1342 ± 0.0400	0.1188 ± 0.0127
2.5%	0.0025 ± 0.0002	0.0041 ± 0.0001	0.1385 ± 0.0213	0.0653 ± 0.0047
Glutathione				
0.5%	0.0180 ± 0.0010	0.0107 ± 0.0004	0.1866 ± 0.0487	0.1993 ± 0.0095
1.5%	0.0094 ± 0.0011	0.0089 ± 0.0004	0.1676 ± 0.0262	0.1311 ± 0.0115
2.5%	0.0061 ± 0.0007	0.0087 ± 0.0004	0.1329 ± 0.0229	0.0973 ± 0.0167

Methods for Detection of Spoilage and Pathogenic Microorganisms on Fruits and Vegetables

Isolation using Plate Count Method

Conventional method of bacterial enumeration works exquisitely on the recovery of viable bacteria from fruits and vegetables. Many food microbiology laboratories lack availability and utilization of novel molecular-based technologies. Due to this, conventional methods such as standard plate count, selective or differential media for isolation and detection and of bacteria samples; and further identification using biochemical methods such as gram stain and other commercially available profiling systems. Non-selective media such as nutrient agar, tryptone soy agar can aid proliferation of bacteria cells at incubation time of 4 h before subsequent transfer to a selective media such as Mac Conkey agar for E. coli, and Salmonella-Shigella agar for Salmonella detection. Bacteria can be further separated into gram positive and negative using gram stain. The process of transfer from non-selective to selective media allows sub-lethally injured bacteria to recover in time for detection. Other biochemical tests used as follow up to conventional isolation of pathogenic bacteria include catalase test for identification of gram positive bacteria and oxidase test for gram negative bacteria.

Use of Fluorogenic and Chromogenic Media

Another media used for isolation and detection of pathogenic bacteria is the use of fluorogenic and chromogenic culture media. Identification of bacteria on fluorogenic media is made possible through the incorporation of enzyme substrates (which consists of a sugar/amino acid-fluorogen complex) into a selective media which in turn speeds up biochemical confirmation of the bacterial identity. For example, methylumbelliferyl is a fluorogen that has been incorporated in an array of media for detection of coliforms such as E. coli 0157:H7 which forms a blue fluorescence upon exposure to ultraviolet light. Rainbow agar, BCM O157:H7 (and other coliforms), CHROMagar, and Colilert are some examples of fluorogenic or chromogenic media for coliforms detection. The most commonly used culture reference methods for the detection of Listeria in foods are the ISO 11290 standards; FDA-BAM method to isolate Listeria spp. from vegetables.

Molecular-based Methods

Nucleic acid-based systems designed for the detection of genomic DNA specific to particular microorganisms are capable of achieving rapidly and highly sensitive identification even when the target microbe is present in low numbers. In order to achieve this, polymerase chain r eaction (PCR) is quite useful. PCR is a molecular-based detection method. This method is focused on extraction of bacteria DNA; and it works best when there is enough bacterial cells from which is boosted by the enrichment step. Enzyme-linked immunosorbent assay (ELISA) is another molecular-based method that works on the principle of antigen-antibody interaction. It is more sensitive and specific to a bacteria strain than standard plate count method. Steps in Antigen (food slurry/extract) is added to sample wells in a microtiter plate containing an antibody with specificity to the target molecule. ELISA has been successfully used to detect virulence determinants of pathogens such as Campylobacter, E. coli 0157:H7 and L. monocytogenes. Serotypes of Listeria spp. were categorized based on specific heat-stable somatic

(O) and heat-labile flagellar (H) antigens. ELISA procedure entails adding sample, washing, adding antibody complexes, adding detection reagents; and these steps are labour intensive. This has led to automation of the ELISA process, thus cutting back on time and labour. A good example is BioMerieux's Vidas System in which the entire procedure is finished in 2 h after addition of overnight enrichment broth. This system is available for assays of pathogenic bacteria such as L. monocytogenes, Salmonella, E. coli, and Campylobacter.

Rapid Methods of Microbial Pathogenic Detection

Standard methods of isolation and identification of pathogenic microorganisms in foods are slower and time consuming which has led to demand of quicker methods; and for the past two decades the latter have been developed for both on-site and laboratory tests. A method can be characterized as rapid when it gives quicker results that the conventional method. Other factors that determine its effectiveness are sensitivity, standardization, reliability, accuracy, specificity, evaluation, ease of use, cost, validation, convenience and potential for automation. Most of the advances in development of rapid methods are in molecular-based methods and other areas including impedance and conductance, bacteriophages, biosensors, microscopy as well as in miniaturized, automated biochemical detection kits. There are currently many diagnostic systems like RapID, Minitek, API, Biolog, MicroID, Crystal ID and VITEK systems which are commercially available for identifying different microorganisms.

Prevention of Contamination, Safety and Hygiene Practices during Processing of FFV

The rate of contamination of FFVP after processing (cutting or wounding) is greater when compared to those of whole fruits and vegetables. This has been largely attributed to high moisture content in the fruits and vegetables as well as wound occurring in the tissues due to processing. Wounding of tissue as a result of slicing, cutting or peeling releases the nutrients inherent in these products thereby enhancing microbial contamination and growth. Upon growth of these microbes on the FFV surface, susceptibility to the formation of biofilms increases in the produce thus bringing about difficulty in the elimination of these microbes. Microbial biofilms are thus able to attach, grow and spread to any surface with the cells associated with the biofilms possessing an advantage in growth and survival over planktonic cells. Growth and survival advantage over planktonic cells by biofilms has been attributed to the formation of exopolysaccharide matrix which surrounds the biofilms, thereby building a wall against the environment and protecting the biofilms from sanitizers.

Pathogens of major concern in FFVP include Listeria monocytogenes, pathogenic Escherichia coli and Salmonella spp. A number of human pathogens have also been implicated in the contamination of FFV with a reported increase in recent years in the number of produce-linked food-borne occurrences. Agricultural practices during harvesting, human handling, quality of water and soil, contaminated equipment, processing methods, use of contaminated packaging materials as well as transportation and distribution have all been implicated in the contamination of fresh-cut produce. Similarly, microbial adhesion on conveyor belts, containers and food contact surfaces used along the food chain has been shown to lead to the formation of biofilms. Microbial contamination has also been shown to lead to internalization of pathogens into the fresh-cut fruit and vegetable produce. E. coli and Salmonella typhimurium are capable of penetrating the leaves of iceberg lettuce. E. coli O157:H7 can penetrate between 20–100 µm below the surface of lettuce leaves. Internalization of such pathogens has been reported to occur in the stomata, vasculature, cut edges and intercellular tissues with elimination of such pathogens rather impossible thus hindering assurance of product safety and rendering processing and preservatory methods completely futile.

Some of the biological methods employed in the reduction of pathogenic attack and spoilage of processed FFV include bacteriocins such as lactic acid bacteria which produces organic acids and bacteriocins that can act as antimicrobials. For instance bacteriocin nisin is a natural preservative produced by Lactococcus lactis and is effective against mostly gram positive bacteria. Nisin acts on the cell membrane of the microbe thereby forming pores that result in cell death during the process. Other biological methods employed include the use of bacteriophages which has found application as disinfectants and preservatives. Other biological methods applied include the use of bacteriophages used in the destruction of bacteria. Bacteriophages are virus that infect bacteria thereby bringing about their death and destruction as a result. Its advantage in its application includes specificity in action; effectiveness; availability and accessibility; and reduced effects on the organoleptic properties of the fresh-cut products.

Enzymes have also been employed in the control of pathogenic organisms in fresh-cut products. enzymes are able to target directly the biofilms that interfere with their development process, speed up the formation of antimicrobials and even destroy mature biofilm. In attacking biofilms, enzymes mostly target the extracellular polymeric matrix which surrounds the biofilm cells and influences the shape of the biofilm structure as well as its resistance to shear forces. Hence, enzymes can be used as an alternative to conventional chemical disinfectants in the removal of biofilms from fruits, leaves and other abiotic surfaces. However, the use of enzymes requires prolonged contact times for effectiveness against biofilms and the fact that extracellular polymeric substances produced by biofilms are mostly heterogeneous confers some disadvantages on the use of enzymes. Accordingly, use of enzymes alone as a biological control against pathogens in fresh produce does not guarantee total removal of biofilms. Lequette et al. and Augustin et al. therefore suggested the use of enzymes in combination with other treatments especially with antimicrobial agents.

Storage of FFV

Modified Atmosphere Packaging (MAP) of Fresh-cut Produce

One of the various ways of processing fresh-cut fruits and vegetable is the use of modified atmosphere packaging (MAP) of the produce. The process of MAP helps in altering the gaseous composition within a food packaging system. MAP relies greatly on the interface between the rate of respiration of the produce and the transfer of gases through the packaging material without any further alteration to the initial gas composition. MAP can either be passive: which involves generation of MAP in a packaging material by reliance wholly on the natural process of respiration of the packaged produce as well as the permeability of the packaging film material in bringing about the desired gas composition. MAP can also be active: involving the replacement of the gaseous composition in a packaged material through the introduction of gas scavengers or absorbers such as ethylene scavengers, oxygen and carbon oxide, thereby establishing the preferred gas mixture within the package.

FFV have a short shelf life due to respiratory metabolism. Modified atmosphere packaging (MAP) has been used to reduce the rate of respiration and water loss leading to prolonged storage period. MAP comprising of low O_2 and elevated CO_2 atmospheres have been used to extend the shelf life and leading to high organoleptic characteristics of pear, apple, mango and peach fresh cut. The effects of low O_2 and CO_2-enriched atmospheres associated to different packaging, traditional and compostable, on shelf life of fresh-cut nectarine slices stored (1 °C) for 7 days by Maghenzani et al. showed that low permeability of the film has a positive influence on weight loss and firmness as the less permeable film allowed a greater water retention, which caused a lower weight loss. They observed that MAP, acting on the respiratory metabolism, reduced respiratory metabolism with positive effect on colour, total soluble solids, titratable acidity, firmness and PPO activity, though efficacy differed among two cultivars of the fruit. Biodegradable films performs better than polyethylene film as a packaging material.

Fresh foods such as fruit and vegetables are alive and continue to respire after harvest. Reducing the respiration rate and reducing the heat produced through efficient airflow inside ventilated packaging is important in maintaining product quality.

Freezing of FFVP

Freezing is a widely known and applied preservation process of various foods which offers the advantage of producing high-quality nutritious foods with prolonged shelf life. Freezing has also been described as one of the best methods used in preserving foods such as fruits and vegetables. Freezing of FFV will reduce the problem of spoilage experienced by the fresh-cut commodities. However, there is a perception by consumers that freezing reduces and affects negatively the nutritional composition of the fruits. A point of comparison is based on the fact that fresh produce could maintain its keeping quality in the consumer's home for a number of days prior to consumption.

During freezing most of the liquid water constituent of the food materials is transformed into ice, thereby reducing water activity, which slows down the physical and biochemical changes involved in the deterioration of foods as well as the growth and reproduction of spoilage microorganisms. Fruits and vegetables are composed of approximately 85–90% water which crystallizes during freezing. The freezing process prevents microbial growth, reduces water activity, and decreases chemical and enzymatic reactions. Decrease in temperature experienced during freezing impedes metabolic reactions taking place in the fruit and vegetable after harvesting. Freezing also reduces the rate of microbiological activities occurring in the FFV positively affecting the overall product quality. During the process of freezing, conversion of water to ice brings about various stress mechanism such as volumetric change of water converting into ice, spatial distribution of ice within the system as well as the size of the ice crystal. The effect of this stress mechanism is the deterioration of the frozen products by affecting the texture and structure of the cut fruit and vegetable.

It is well known that FFV undergo faster physiological deterioration, biochemical changes and microbial degradation which may result in degradation of the colour, texture and flavour. However, the high water content of fresh-cut products adversely affects the textural quality of the products after thawing due to the formation of ice crystals and water solids within the cell structure. When thawing takes place the cellular structure of the fruit and vegetables is destroyed. The reduction in the product water content results in improved freezing performance and ameliorated product quality including better preservation of structural and textural characteristics. Thus, in order to preserve the structural and textural characteristics and improve freezing performance, the water content of the fruits and vegetables are reduced by dehydration before freezing. Frozen fruits and vegetables are mostly consumed cooked with majority of vegetables blanched prior to freezing. Blanching action prior to freezing has been reported to influence greatly the structure of the vegetable thus resulting in an initial loss of firmness due to disruption of the plasma lemma and an increase in the ease of cell separation accompanied by swelling of the cell wall.

Several novel freezing practices are presently being investigated to overcome the problems of FFV and other food produce undergoing physical and chemical changes as a result of freezing. One of such novel methods which is presently being explored is dehydrofreezing. During dehydrofreezing process, the food is first dried up to a needed moisture content level before the onset of freezing. Hence it is aptly described as a process of freezing relatively dehydrated foods. For fresh-cut products, non-thermal dehydration techniques such as vacuum and air drying are mildly applied prior to freezing. When the method of drying the FFV is through osmosis then it is termed osmodehydrofreezing. Dehydrofreezing is particularly well suited for fresh-cut fruits and vegetables due to the fact that reducing the moisture content in the produce will allow for the formation and expansion of ice crystals without damaging the cellular structure of the product. Theoretically, the dehydration treatment not only reduces the amount of water to be frozen but also makes cell structures less susceptible to breakdown by changing cell turgor pressure. The reduced water content has the potential to reduce the freezing time, the initial freezing point and amount of ice formed within the product. As a consequence, the damage to plant cells caused by ice crystal formation and the post-thawing quality degradation such as softening and loss of good textural attributes are alleviated. Reduction in moisture content before freezing also leads to reduced freezing time since there is less water to freeze as well as a reduction in the amount of ice formed within the produce. Fruits and vegetables are said to exhibit better quality over those that are frozen without any form of reduction in moisture content.

Generally, the texture of a thawed frozen fresh-cut fruit and vegetable product is much softer than normal produce due to cell rupturing caused by expansion of the plant cells during freezing. Hence the recommendation that the moisture content of fresh-cut produce be reduced before freezing in order to mitigate the effect of freezing on the thawed product.

Nutritional Content of FFV

FFV are derived from whole fruits by cutting them into desired shapes and sizes. However, peeling and cutting cause serious damage to vegetable tissues which leads to dissociation of cell components that brings about biochemical reactions such as accelerated oxidative browning and chlorophyll degradation. Other quality deterioration include water loss, development of off-flavours, stimulation of microbial growth and tissue softening which makes fresh-cut fruits have short shelf life. Wounding stress as a result of cutting first causes the plasma membrane to break thereby inducing reaction of oxidative enzymes with existing phenolic compounds causing oxidation of the latter.

Nutritional value of fresh-cut fruits is usually a measure of vitamins A, B, C, E, polyphenolics and carotenoids; while that of vegetables include the vitamins, glucosinolates, carotenoids and polyphenolics through spectrophotometric and colorimetric methods. Li et al. tested the effect of cutting a whole pitaya fruit into slice, half and quarter slices on its nutritional quality. Their results revealed that the various cutting styles had little influence on vitamin C and soluble solids. However, total phenolic content, antioxidant activity, increased significantly with cutting wounding intensity up to first two days of storage before deterioration set in. Some nutritional contents of selected fruits and vegetables are highlighted in table below.

Table: Nutritional content of some fruits and vegetables.

Fruits/veg	Sugars (°Brix)	Vitamin C (mg/100 g)	Polyphenols (mg/100 g)	Antioxidants (%DPPH)
Kiwi	12.6	44.99	-	-
Apple	-	3.00	329.30–660.00	12
Pitaya	0.15 g/kg	0.16 g/kg	0.6–1.2 g/kg	40–70
Papaya	11.82	13.3–18.0	1.9–3.5	-
Pineapple	-	76.27–142.41	42.28–46.70	30–60
Banana	-	46.75–53.25	60.25–85.24	70–90
Guava	-	65.16–198.25	96.51–178.51	-
Radicchio leaves	-	-	143–360	-
Swede	535.2 g/kg	-	-	-
Turnip	468.9 g/kg	-	-	-

Sugars

Total sugar content of swede were not affected by storage temperature; while lower temperatures (0 & −2 °C) increased the sugar content of turnip than higher temperatures as a result of glucose and fructose metabolism by enzymes at lower temperatures. Benítez et al. reported that soluble solid content of kiwi slices coated with aloe vera gel, alginate and chitosan did not significantly change up till day 8 of storage at 5 °C.

Vitamin C

Vitamin C is the vitamin that usually degrades most rapidly and can be used as an index of freshness. Vitamin C is unstable in many vegetables such as asparagus. There was no significant difference in the ascorbic acid contents of FC papaya stored at 10–20 °C for 0 to 24 h but a significant decrease was observed at storage temperature and time of 4C, 48 h. There was no significant influence of cutting style on vitamin C content of pitaya fruit when cut in slice, half and quarter shape. Exposure of fresh-cut banana, pineapple and guava slices to ozone for 0–30 min drastically reduced vitamin C contents of the fruits by 12.21, 46.44 and 67.13% respectively. Vitamin C content of kiwi slices coated with chitosan and alginate, stored at 5 °C depreciated at storage time from day 1 to day 11. However, kiwi slices coated with aloe vera gel significantly increased from 44.99 to 47.99 mg/100 g at the same storage conditions.

Polyphenols

When wounding stress occurs, plants produce injury signals to induce the production of more secondary metabolites including phenolic antioxidants to defense and heal the wounding damage. Wounding stress also activates phenylalanine ammonia lyase (PAL) - an enzyme responsible for synthesis of phenolic compounds in plant tissues. For example, carrots synthesize lignin along wound barriers. Flavonoid contents of fresh-cut papaya significantly increased at storage conditions of 20 °C at 24, 36 and 48 h when treated with 405 nm LED illumination. Li et al. reported a gradual increase in total polyphenol content of pitaya fruit with storage time of 4 days. An approximate increase of 63, 78 and 90% was reported for slice, half and quarter slice respectively at day 2; after which a decline was observed. Total polyphenols and flavonoids in fresh-cut pineapple, and banana increased as the fruit slices were exposed to ozone for up to 20 min; but a reverse trend was observed for guava slices. The reason for increase polyphenols by ozone treatment was attributed to activation of PAL. Upon 1 day in storage, total phenolic content of untreated and fresh-cut apples treated with citric acid and UV light decreased by 50%.

Antioxidants

Kim et al. observed no significant difference in antioxidant capacity of fresh-cut papaya stored at 4–20 °C for 0 to 48 h. Antioxidant capacity of fresh-cut pineapple and banana increased when exposed to ozone for 20 min and declined upon further treatment; while that of fresh-cut guava reduced with ozone treatment and increased when ozone exposure time was increased to 30 min. A drastic loss of antioxidant potential of fresh-cut apples was observed in both untreated and fruit pieces treated with pulsed light and gellan-gum coatings during the first week of storage.

POSTHARVEST HANDLING OF APPLE

Harvesting at the correct time is essential to the production of quality apples. To ensure maximum storage life, apples should be harvested when mature but not yet fully ripe or overripe. If harvested before they have matured, apples will have poor eating quality, will be more susceptible to storage disorders such as scald, cork spot, and bitter pit, and may not ripen properly. Ripe fruit should be avoided because it will continue to ripen in storage, rapidly becoming too soft and mealy for sale. Firmness and the level of soluble solids in the apple are good indicators of maturity to use in determining picking time.

Apples are very susceptible to bruising and other forms of mechanical damage and therefore should not be handled any more than necessary. Workers harvesting apples should be cautioned not to drop them or handle them roughly. The effects of bruising and scuffing cannot be reversed. Damage from rough handling will accelerate deterioration, reducing the value of the product.

Apples are normally transported and stored in bulk boxes filled in the orchard. Each of these boxes holds about 600 pounds of fruit. Full boxes should not be allowed to sit for extended periods in direct sunlight nor for more than a few hours before cooling is started. They also should not be overfilled. When overfilled boxes are stacked, many apples throughout the box are bruised.

Bulk boxes represent a substantial investment for most growers and processors. They should be kept clean and in good repair and should be stored indoors when not in use. Most bulk boxes made from hardwood have enough open space between the individual boards in the bottom and sides to ensure adequate air circulation and water drainage. For some of the newer boxes built from plywood, it may be necessary to drill holes to improve air circulation. Five to 8 percent of the lateral surfaces and 3 to 5 percent of the bottom should be open. Better circulation is provided by many small, well-distributed holes than a few large ones.

Chemical Treatments

Harvested apples are often treated with chemicals to inhibit storage disorders. Dip or spray treatments are commonly used to prevent scald and a group of disorders such as bitter pit that are associated with low calcium levels in the fruit. A good aqueous solution is made of 3 percent calcium chloride plus a scald inhibitor mixed with a surfactant and fungicides.

A. 0.1 percent concentration of surfactant is usually sufficient to provide complete wetting of the fruit, but more surfactant can be added if droplets begin to form on the apples. If the solution is applied to fruit in bulk boxes or bushel boxes, the calcium chloride may severely corrode the metal fasteners after a few seasons. Rinsing the empty containers with water will help alleviate this problem.

Cooling and Storage

An apple continues to live and respire even after it is picked. Although respiration cannot be halted completely, the objective of postharvest cooling is to slow the process and thus increase storage life. Even if apples are to be stored for only a short period, it is still very important that the field heat be removed from them as soon as possible. Rapid cooling will not injure the apples. They may be either hydrocooled or forced-air cooled without removing them from the bulk boxes. If they are hydrocooled, they can also be drenched with a scald inhibitor and fungicide in the same operation.

The higher the holding temperature, the greater the softening and respiration rate and the sooner the quality become unacceptable. Apples respire and degrade twice as fast at 40 °F as at 32 °F. At 60 °F they will respire and degrade more than six times faster. The optimum storage temperature for apples depends on the variety, but all are within the range from 30 to 40 °F. Most apple varieties grown in North Carolina are best stored at or near 32 °F. Apples are moderately susceptible to freeze damage. Temperatures more than 1 or 2 degrees below freezing should be avoided.

All varieties require a relative humidity from 90 to 95 percent, which may require adding water vapor to the air in the storage room with one or more humidifiers. Maintaining the humidity within this range will reduce weight loss, but humidity near the saturation point (100 percent) will encourage the growth of bacteria and fungi. Some varieties of apples, such as Golden Delicious, can be held in plastic liners to keep the humidity at high levels.

Chilled apples that are suddenly transferred into warm air are likely to "sweat" - that is, water vapor in the air will condense on them. Sweating also occurs when the doors of a cold storage room are opened, allowing warm, moist air to enter. Sweating itself does not harm the fruit, but it causes wetting, which encourages the growth of fungi and bacteria. Chilled apples should not be allowed to warm and then be rechilled. To prevent sweating, allow chilled apples to warm gradually inside the storage area, if possible, before bringing them out into the open air. Since apples are stored longer than many other types of produce, it is essential that both the storage rooms and containers be clean and sanitary. Storage rooms should be thoroughly cleaned before they are filled. If molds are found to be growing in the storage room, the interior surfaces may be disinfected with a 0.25 percent solution of sodium hypochlorite (1 gallons of household chlorine bleach in 20 gallons of water) applied with a high-pressure washer. Surfaces should be allowed to air dry for several days.

Refrigeration coils, fans, and ducts should also be inspected and cleaned regularly. Refrigeration coils are especially likely to become clogged with dust and dirt that will substantially decrease their thermal efficiency. Thermostats and wet-bulb thermometers should be recalibrated from time to time with an accurate mercury thermometer. Humidistats can be checked for accuracy with a sling psychrometer. If possible, avoid positioning the sensing elements of controllers on exterior walls or on the ceiling. Instead, locate them in the open near the apples.

Remember that thermometers measure the temperature of the air and not the fruit. Fruit temperature will almost always be higher than air temperature and can be accurately measured only with several pulp thermometers inserted into the fruit in various locations throughout the room.

It is also a good idea to check all interior and exterior surfaces periodically for air leaks and damaged insulation. Door gaskets should be checked for a good seal by inserting a sheet of paper between the door and the frame, closing the door, and then attempting to pull the paper out. If it slips out easily, the gasket should be replaced or adjusted.

Cooling Methods

To reduce the cooling load and make the cooling process as energy efficient as possible, apple temperatures should be as low as possible before cooling begins. As much fruit as possible should be harvested during the cool hours of the day to reduce the cooling load. Allowing the apples to sit outside overnight in bulk boxes or trucks will generally not lower their temperature much and may actually cause the apple pulp temperature to increase. Several methods are satisfactory for cooling apples. Figure shows relative cooling times for the three most common methods, described in the following sections. Whichever method is used, care should be taken to ensure that the apples cool quickly and thoroughly with the least use of labor and energy.

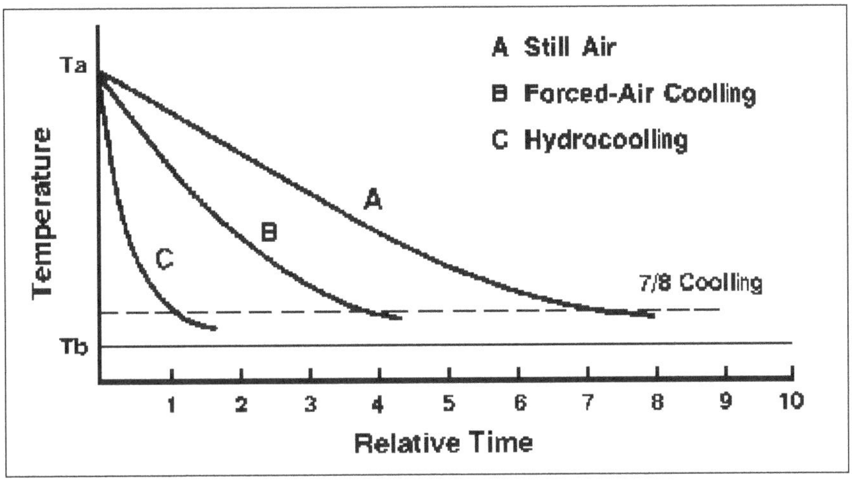

Rate of temperature change for three cooling methods.

Room Cooling

For many years apples have been cooled by storing them in refrigerated rooms. One of the least expensive methods for cooling apples, room cooling is accomplished by simply stacking bulk boxes inside a refrigerated room where the heat is allowed to dissipate slowly. This method requires a minimum of handling and labor. After cooling is completed, the facility can be used for short-term storage as well. Because cooling occurs over a period of several days or even longer, the electrical energy demand from the refrigeration system is relatively low. As a result, a smaller, less expensive cooling system can be used, and the cost of electricity is lower than for more rapid cooling methods.

A disadvantage of this method is that it may take too long to cool the apples. Tests have shown that when apples in tightly stacked bin boxes are placed in a cold storage room, it may take from several days to more than two weeks for the apples to reach the same temperature as the air. In some cases the rate of heat transfer from apples in the middle of the box is insufficient to overcome the temperature rise produced by natural respiration. Without forced air movement, heat is lost only from the produce near the outside of the package. As a result, cooling is uneven and the cooling time may be too long to prevent serious deterioration and loss of quality.

If room cooling is used, be sure to allow adequate space for proper air circulation between rows of stacked bulk boxes. Do not stack rows of bulk boxes less than 6 inches apart nor less than 8 inches from an outside wall or ceiling. Properly directing the flow of air from the refrigeration coils can significantly increase the rate of cooling.

Forced-air Cooling

It is becoming more common to use forced-air cooling to cool apples quickly. Forced-air cooling is accomplished by exposing the bulk boxes in a storage room to a higher air pressure on one side than the other. The unequal air pressure forces the cool air past the produce, greatly increasing the cooling rate. It is essential, therefore, that the apple containers have sufficient open space to allow for air movement. Depending on the fan capacity and the produce being cooled, forced-air cooling is from 4 to 10 times faster than room cooling. Fan selection is critical. Not all fans can produce enough pressure to move air through the produce. Forced-air cooling fan systems can be designed to meet various cooling load requirements. They should be able to operate efficiently under relatively high static loads and to deliver about 2 to 3 cubic feet of air per minute for each pound of fruit. One sufficiently large fan can be used to cool as many as 50 standard bulk boxes of apples.

Several produce package arrangements are acceptable for forced-air cooling. In general, containers are arranged in a shell fashion. Air is always pulled, never blown, through no more than one layer of apple boxes. The faster the air moves past the apples, the faster the apples cool, but also the faster they lose water. Therefore, the forced-air cooling system should be operated only until the apple pulp temperature reaches 38 to 40 °F. Forced-air cooling requires close monitoring of the relative humidity. If the relative humidity dips below the optimum 90 to 95 percent level for even a short period, substantial water loss and fruit shrinkage will occur.

Forced-air cooling lowers the temperature of the apples much more rapidly than room cooling. A larger refrigeration system may be required to handle the increased load, a fact to be considered when planning a new facility or when switching to forced-air cooling. Determinations need to be made on an individual basis. Even though the electrical energy demand is higher for forced-air cooling, a properly designed system can increase overall cooling efficiency by 15 to 25 percent because it takes less time to cool the fruit.

Hydrocooling

A hydrocooling system.

One of the quickest methods for removing field heat from apples is hydrocooling. This process is accomplished by flooding the fruit with large volumes of chilled water, normally in a hydrocooler designed specifically for that purpose. Washers used to apply fungicides and scald inhibitors usually cannot be successfully converted into hydrocoolers. It may be possible, however, to apply fungicides and scald inhibitors during hydrocooling. If a sufficiently large quantity of cold well or stream water is available, it can be used as a relatively low-cost source of hydrocooling water. Stream water should be used only after it has been checked for purity. For hydrocooling to be quick and efficient, the water must:

- Be kept at a temperature as near to 32 °F as possible.
- Move past the apples as fast as possible.
- Contact all the apples.

It is not advisable to remove all the heat by hydrocooling alone. To save time and energy, apples are seldom hydrocooled to lower than 45 °F. Slower methods such as forced-air or room cooling are usually employed to remove the balance of the heat. The rate at which apples may be hydrocooled depends on their size. In general, as the diameter of the apple doubles, so does the length of time required to cool it. This means that large apples need to be hydrocooled longer than small ones. However, pound for pound, the same amount of cooling capacity is required. An approximate rule of thumb to use in determining the size of the refrigeration systems needed for hydrocooling is to provide 5 tons of refrigeration capacity for each thousand pounds of apples cooled per hour.

POSTHARVEST HANDLING OF LICHI

Once harvested, exposure of litchi fruits to the sun and air can increase water loss by many folds. Therefore, harvested fruits can be kept at cooler place in the orchard or in temporary thatched house for cooling. A tarpaulin can be used to protect the fruit, but must be kept clean to prevent the build-up of pathogens. Spraying the fruit with water may help to maintain fruit quality in hot, dry weather. The transfer of fruit to the pack house soon after harvest minimizes the opportunity for water loss in the field. Transporting the fruit dry and fairly tightly packed reduces the risk of vibration damage.

Sorting of produce can be done during the harvest operation itself. Scarred, cracked, discoloured or otherwise affected fruits can be placed in separate harvest containers or discarded completely and removed from the field later. Precise operations such as washing and grading according to critical size, colour, etc. may be better performed in a pack house. There is a tendency/practice of packing produce directly into final marketing containers in the field, reducing the number of times the produce is handled with a corresponding reduction in the amount of injury caused but this requires appropriate grading, precooling and packaging facilities and adequate supervision.

Collection and bringing harvested produce together for the purpose of further process is necessary. Field assembly should be planned bearing in mind the best location and the provision of basic facilities. Containers filled with produce should not be left exposed to sunlight. As with all aspects of post-harvest, the objective of assembly should be to develop a system which minimizes stress on the produce and keeps handling operations to a minimum but with maximum care and to keep the time between harvest and first destination of the produce as brief as possible.

Sorting and Grading

To get the produce from big orchard to a collection point by the roadside may involve passing over one or more kilometers of farm roads. Growers rarely give sufficient attention to the logistics of this operation but it should be done properly. Specific fruit quality is maintained by removing damaged and inferior fruit during sorting. Close attention to all the fruits under good lighting is required at this stage. The entire surface of each fruit must be observed to ensure that damaged specimens are not packed. Damage extending to the aril rapidly leads to rots, which may spread to sound fruit within the package. For this reason, fruit with pulled stems, splits, cracks and insect damage should be rejected at this stage. Immature fruit and fruit showing any signs of rot are also removed during sorting.

Finally, the fruits are graded according to their size, weight, colour and maturity. Grading separates fruit into different grades to suit different markets. Most producers have at least two grades of fruit. Grading is normally carried out during or after sorting. Grading systems depend on market requirements, but are normally based on fruit size and colour, and the area of blemish. Export markets usually have higher standards than domestic markets, requiring uniform, unblemished fruit. There can also be differences within different sections of the domestic market. As per fruits and vegetables grading and marking rules, the grade designation and quality of litchi for export will be:

- Litchi shall be:
    - Whole, sound, fresh in appearance.
    - Clean, free from any visible foreign matter.
    - Free from pests affecting the general appearance of the produce.
    - Free from damage caused by pests.
    - Free from abnormal external moisture excluding condensation following removal from cold storage.
    - Free from any foreign smell and/or taste.
    - Free from damage and abrasion.
    - Free from brown markings.

- Litchis shall have minimum equatorial diameter of 23 mm.
- Litchis shall comply with the residue levels of heavy metals, pesticides and other food safety parameters as laid down by the Codex Alimentarius Commission for exports. Litchis shall be presented under one of the following forms:
    - Individually: In this case the pedicel must be cut at the first knot and the maximum length of the stalk must not extend more than 2 mm beyond the top of the fruit. Litchi in special grade must be presented individually.
    - In bunches: In this case the bunch must include more than three attached and well-formed litchis. The branch must not exceed 15 cm in length.

Pre-cooling

To remove the field heat litchi fruit should be hydro-cooled or air-cooled. Pre-cooling requirement of a fruit is largely determined by its physiology in relation to harvest maturity and ambient temperature at harvest time. Highly perishable litchi must be pre-cooled before storage, as the ambient temperature at the time of harvest in India varies from 25-30 °C with 40-45% RH. This is highly detrimental for storage of litchi fruits causing rapid desiccation. There are two methods for pre-cooling of fruits after harvest i.e. hydro-cooling and air-cooling. Cooling speed is generally greater in hydro-cooling system without dehydration than aircooling. When hydro-cooled, the internal temperature of the fruit is reduced to 5 °C within 18 minutes, whereas air-cooling takes longer time. Hydro-cooled fruits packed in polyethylenelines boxes remain fresh and give the best appearance after 6 days of storage at 5 °C. The precooling treatment, however, has also been found advantageous even for storage at relatively higher temperature. Pre-cooled litchi fruits, when treated with 0.5% lectithin and 2.5% sodium bicarbonate and packed in 0.04 mm polythylene bags, retain their fresh colour upto 8 days of storage at room temperature (25 °C). Hydro-cooling of litchi at 0-2 °C and maintenance of a cold chain throughout its post harvest life i.e. from picking to retailers shop has been found useful. Fruit which are to be sold in local markets should not be refrigerated, since on subsequent exposure to heat or humid air rotting will quickly set in.

Table: Criteria for grade designation.

Grade designation	Grade Requirements	Grade tolerances
Extra class (Equatorial diameter 33 mm)	Litchis must be of superior quality. They must have the shape, development and colouring that are typical of the variety and/or varietal type. They must be free of defects, with the exception of very slight superficial defects, provided these do not affect the general appearance of the produce, the quality, the keeping quality and presentation in the package.	5% by number or weight of litchi not satisfying the requirements for the grade, but meeting those of Class I grade or exceptionally coming within the tolerances of that grade.

Class I (Equatorial diameter 28 mm)	Litchis must be of good quality. They must be characteristic of the variety and/or commercial type. The following slight defects however may be allowed provided these do not affect the general appearance of the produce, the quality, the keeping quality and presentation in the package: • Slight defects in shape, • Slight defects in colouring, • Slight skin defects, • Provided these do not exceed a total area of 0.25 sq.cm.	10% by number or weight of litchis not satisfying the requirements of the grade, but meeting those of Class II grade or, exceptionally coming within the tolerances of that grade.
Class II (Equatorial diameter 23 mm)	This grade includes litchis which do not qualify for inclusion in the higher grades, but satisfy the minimum requirements specified in general characteristics. The following defects may be allowed, provided, the litchis retain their essential characteristics as regards the quality, the keeping quality and presentation: • Defects in shape. • Defects in colouring. • Skin blemishes provided these do not exceed a total area of 0.5 sq. cm.	10% by number or weight of Litchis not satisfying the requirements of the grade, but meeting the minimum requirements.

Packaging and Transport

Sorting, grading and packing are often carried out in a pack house or shed, to protect workers and fruit from the heat. Where shelter is not available, operations are best located in a cool, shady area. Where temperature should not be more that 12-15 °C. Good hygiene in the pack house is required to avoid the spread of diseases during handling. Pathogens can build-up on packing surfaces and fruit crates. These surfaces should be washed with sanitizing agents such as chlorine every day. Water and fungicide dips also require frequent replacement or sanitizing. Waste fruit need to be regularly removed from the packing area to reduce the spread of spores. Proper packing, labeling and safe transport are to be done for good revenue.

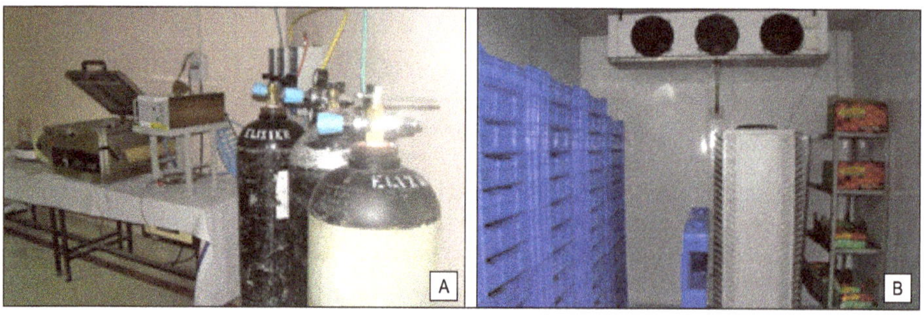

After harvesting, fruit should be packed as quickly as possible, as their quality deteriorates markedly if they are exposed to sun even for a few hours. Litchi can be packed in panicles, or as individual fruit. Fruit are often sold on panicles in Asia, whereas loose fruit are more common in Australia, Europe and North America. De-stalking is required when fruit harvested on panicles and packed individually. The first step in the packing process consists of cutting but allowing a portion of the pedicel to each fruit. All the cracked or damaged fruits must be removed, because such fruit ferment within 1 or 2 days due to bacterial activity, thus creating favourable conditions for a fungus attack, which could then spread to healthy fruits.

Browning in litchi fruit occurs within 24 hours at ambient temperatures of 20-30 °C due to enzyme activity, probably an aerobic system of oxidative poly phenolase which is accelerated by drying. Two conditions, humidity and low temperatures are vital to prevent darkening and to preserve the taste qualities of the fruit. This was a well-known fact in ancient China where it was customary to spray the fruits with a solution of cold salt water and then pack them in bamboo containers sealed with clay. It is worth noting that while low temperatures help to prevent rot from developing, high humidity has a negative effect and care should be taken to ensure that humidity is not excessive.

Selection of packaging depends on market preferences and availability. The ideal package protects fruit from damage and minimizes water loss and condensation. For domestic markets litchi is usually packed in small bamboo baskets or wooden crates. Much of the fruit marketed in Asia is transported in bamboo baskets. Square baskets less than 30 cm high give good protection against injury. Circular baskets can be improved by tying strings in two or three directions across the top, with padding underneath. The outer layers of fruit in these baskets are prone to rapid water loss. This can be alleviated by lining the baskets or by covering them with a tarpaulin. Many of the larger commercial operations now consign their fruit in plastic trays or cardboard boxes, which provide better control of water loss. Some growers in Asia dip their fruit in cold water or cover their fruit with ice. However, very low temperatures can injure the fruit. Free water around the fruit as the ice melts can also increase the risk of diseases.

Proper packing of fruits is important in maintaining freshness and quality and preventing fruit decay during transit for marketing to distant places. These are lined with litchi leaves or other soft packing material as paper savings, wood-wool, etc. Sissum leaves are supposed to be the best material for packing. A good box for packing fruits should be light in weight, shallow and rigid enough to protect the fruits. It should have few holes for ventilation and rope handles on either side for lifting the box. Fruits are packed in clusters along with few leaves. Plastic packaging combined with cool storage or fungicides has also been recommended. However, without good temperature control, plastic covers result in condensation and an increased risk of rots.

Labeling

Growers packing litchi for sale in the domestic markets must mark on the top of the box the following information:

- The name of the grower or packing establishment.
- Street/road and area, not the postal address.
- The litchi variety, class and count if packed in litchi cartons.
- Date on which fruit was packed.
- Fruit in bulk packs must have the weight shown instead of count number.

Transport

Poor transport conditions are a major problem of litchi. The main limitations, including rough roads, lack of refrigeration and poor truck suspension which are out of the control of growers. However, with the advancement of transport facilities and road connectivity, the situation has improved in almost all the litchi growing states, the huge quantities of fruit are transported from distant producing centers mainly by road and sometimes by rail. For local market the harvested fruits are packed and transported by trucks. While transporting litchi, following care should be taken:

- Packing between vehicles and baskets with soft straw or leaves can help cushion the blow but it should be clean. Shelves can be used in the truck to avoid stacking of baskets, and reduce damage to fruit. Padding and strapping the baskets can restrict movement during transport. Vehicles should be driven slowly. Do not ride on top of the produce.
- Fruit are often damaged when baskets are overfilled, dropped, stacked or packed on their sides. Vehicle should be shaded from both sun and rain.
- Fruit can be protected by covering the baskets with a clean tarpaulin or similar material. However, it should not prevent ventilation of the produce.
- Often the above points are overlooked because the operators argue that the produce does not have far to travel and the produce is not expected to be on the vehicle for long. Unfortunately, drivers do have the habit of making unscheduled stops during their journeys. They may stop due to road blockage or due to personal work and often the Van is parked in the full sun.
- Exposure to warm air can dry out the fruit very quickly, so transport during the warmer part of the day is best avoided, if possible.
- For better quality fruits of litchi cool chain should be maintained.

The major concern to traders and growers is the exorbitantly high rates charged by truck operators for transporting litchi. Taking advantage of the very short shelf life of litchi, truck operators are demanding double the normal freight charge. The railways also provide facilities for transporting litchi by attaching bogies on a daily basis. However, the desired impact of this initiative is yet to be felt at the field level.

Storage

Storage is essential for extending the consumption period of fruits, regulating their supply to the market and also for transportation to long distances. Factors affecting storage life are, post harvest disease infection, pressure damage caused by over filling packages, package failure due to over stacking (collapse of bottom layers) and moist cartons losing compression strength when stacked. Litchi fruit cannot be kept for more than a few days after harvest, at room temperature. If marketing is delayed, fruits should be kept in cold storage to avoid rapid loss of colour and quality. Pre-cooling of fruits should be done before cold storage because without pre-cooling, browning could be more rapid compared to the storage of pre-cooled fruits at 2-5 °C. It brings not only appreciable reduction in the rate of ageing and water loss in the fruits, but also makes litchi hardy on prolonged cold storage.

The cold storage without pre-cooling and chemical treatment has less impact on colour maintenance. Pre-cooling as well as chemical treatment has recorded shelf life upto 27 days at 5 °C. Fruits treated with 2% sodium hypochlorite can be stored satisfactorily in perforated polythene bags at 0-3 °C for 25 days. For short-term storage less than two weeks, a temperature of 7 °C is satisfactory. A relative humidity should be kept at 90-95% throughout storage and transport. Controlled atmosphere storage (3-5% O_2 and 3-5% CO_2) reduces skin browning and slows down the losses of ascorbic acid, acidity, and soluble solids. Exposure to oxygen levels below 1% and/or carbon dioxide levels above 15% may induce off-flavors and dull gray appearance of the pulp.

POSTHARVEST HANDLING OF MANGO

Mango (Mangifera indica L.) is one of the most important and popular fruits and is referred to as the "king of fruits" because of its excellent overall eating characteristics. It is consumed as a fresh fruit, in the frozen, preserved or dried forms or is processed into juices, purees, chutneys and pickles. Ripe mangoes are best eaten as fresh fruit, usually as a dessert and are used in the production of confectionery, ice cream, and bakery products. Mango contains a variety of phytochemicals and nutrients. The fruit pulp is high in dietary fiber, Vitamin C, provitamin A, carotenoids and diverse polyphenols.

Examples of mango varieties grown.

Importance of Postharvest Handling

Losses in quantity and quality (or post-harvest losses) occur after harvest at different points in the handling chain. A loss assessment study conducted under this project, showed that losses amounted to 31 percent due mainly to decay that manifested at the retail level. Bruising and weight loss which represent loss in marketable weight are the other nature of losses. These losses can be attributed to poor harvesting, rough handling, and poor packaging and transport conditions. The high incidence of decay mainly in the form of anthracnose and stem end rot can be attributed to the high level of pre-harvest infection due to improper disease management during production.

Decay of mangoes at the retail level.

Anthracnose is a latent infection and symptoms of the disease become apparent only as the fruit ripens. Post-harvest losses represent a waste of the resources – land, labour, energy, water, fertilizer, etc. that went into producing the crop. Every effort must therefore, be made to minimize these losses. Unless these losses are minimized, the gains from production will be offset and potential income cannot be realized. Moreover, consumers are now increasingly paying attention to quality and safety. They are looking for and are willing to pay a premium price for good quality mangoes that are safe. Preservation of the nutritive value goes hand in hand with the prevention of quality deterioration.

Good quality mangoes desired by consumers.

With the changing tastes and lifestyles of consumers, and the need to reduce the high levels of post-harvest losses, the need for good post-harvest handling becomes a matter of concern. The continuous growth of supermarkets and hypermarkets and the demands from institutional buyers require giving more attention to post-harvest handling of mango to satisfy demand for better quality and safe produce.

Characteristics of Mango Fruit that Impact on Postharvest Life

Mango Fruits Undergo Programmed Changes

Stages of ripeness of mango fruit.

The mango fruit, being a living tissue is subject to continuous change until it completely deteriorates. The process of ageing and eventual dying in living tissues is called senescence. There are programmed changes in the fruit like ripening that influence appearance, flavor, texture and nutritive value that cause them to age. While some changes are desirable (e.g., changes associated with ripening), most are deteriorative. These post-harvest changes cannot be stopped but can be slowed down within certain limits. In general, the application of post-harvest technology can only maintain, not improve or little if any, the quality of harvested fruit. The main objective of applying any post-harvest technology is to keep fruit quality and safety as high as possible until it reaches the final consumer.

Mangoes Lose Water

The pulp of the mango fruit consists mainly of water. When harvested, the fruit can no longer replace the water that is lost through respiration. Mangoes are, therefore, subject to shriveling and weight loss and consequently loss in marketable weight. Exposure of the fruit to the sun results in rapid water loss.

Exposure of fruit to the sun leads to shriveling and loss in marketable weight.

Mangoes are Prone to Decay

Bamboo baskets as field containers.

Fruits are susceptible to attack by insects and decay-causing organisms (called pathogens) which can eventually accelerate their deterioration. Rough handling may create wounds which serve as entry points for pathogens. Field containers like bamboo baskets have rough and pointed edges that could damage the fruit, hence the need to line the containers with clean plastic sacks or jute sacks to prevent injury on the fruit.

Mangoes are Prone to Injury

Mangoes are susceptible to damage. When damaged, all of the biological processes within the fruit, like respiration and ethylene (the ripening agent produced naturally by mangoes) production proceed at a rapid rate leading to rapid deterioration. Moreover, damage may not be apparent externally but can show up later in the handling chain typified by the failure of the damaged portion to change color and the presence of a starchy layer just beneath the peel of the ripe fruit.

Rough handling leads to damaged portion with white starchy areas and failure to turn yellow.

Harvest Management Practices

A fruit is considered mature when it has reached a stage of development such that after harvesting and during post-harvest handling, both the external and internal quality attributes are acceptable to consumers. Harvesting at the proper stage of maturity, the manner in which the fruit is detached from the tree and the time of harvesting are important considerations during harvesting.

Maturity Indices

Mango fruits harvested at the correct stage of maturity develop good peel and pulp color and have full flavor and aroma at the ripe stage. Mangoes harvested at an immature stage of development can be induced to ripen but the quality of the ripe fruit and particularly the flavor, is inferior.

Fullness of cheeks, (a) flat shoulder at stem end and (b) presence of bloom as indices of maturity.

Maturity indices are an indication of the readiness of the fruit for harvest. There are several methods of assessing the maturity of mangoes. These include:

- Shape: Flat shoulder at stem end; fullness of cheeks.
- Peel appearance: Presence of bloom; change in peel color from dark to light green (for some varieties).
- Pulp color: Light creamy yellow pulp.

Time of Harvesting

The recommended time for harvesting mature fruits is from 9:00 a.m. to 3:00 p.m. to minimize latex flow. Latex stain which leads to unsightly latex burns are major causes of rejection especially if fruits are to sold on export and premium markets like supermarkets and hotels. Harvesting immediately following rain, should be avoided.

Latex stains in mango due to improper time of harvesting.

Harvesting Method

Mango fruits are generally harvested manually using a picking pole with net bag attached for catching the fruit once it is severed from the branch. The harvested mangoes are then transferred to a collecting container that is lined with clean plastic or jute sack to prevent damage on the fruit. Plastic crates are the most suitable field containers. Harvesting tools and aids such as field containers must be clean to avoid contamination.

(a) Recommended mango picking tool with net at the end and (b) plastic crates as field containers.

Minimizing Damage and Deterioration during Harvesting and Field Handling

- Harvesters must wash their hands before harvesting. Good personal hygiene must be practiced to avoid cross-contamination of the produce.
- Line collecting baskets with clean plastic sacks or newspaper to protect fruits from punctures and cuts caused by sharp or protruding edges of the baskets. Cuts and punctures provide avenues for water loss and invasion of decay-causing organisms. Wounded fruits also produce a lot of ethylene that hastens the ripening process.
- Gently lower filled collecting baskets to the ground. Do not pour harvested fruits onto the ground – pour on to a clean tarpaulin. Harvested mangoes should not get in contact with the soil in order to avoid microbial contamination.
- Minimize damage and avoid contamination of fruits with soil, pathogens, fertilizers or other agro-chemicals.
- Avoid exposing harvested fruits to the heat of the sun to prevent heat buildup.
- Leave 2-3 cm of the stem (pedicel) attached to divert latex flow away from the fruit.

Bad practices during harvesting and field handling.

Post-harvest Operations

Typical packinghouse for mango.

Post-harvest operations refer to activities done to the fresh produce to meet the requirements of the target market. These operations can be done in the field, in collection centers or in a packinghouse. The packing area should provide adequate protection from sun and rain, kept clean at all times, pet animals should be kept away from the packing area, and all workers should practice good personal hygiene. The sequence of operations should insure that cross contamination is avoided. The surroundings of the packinghouse should always be kept clean.

Trimming

Trimming refers to the cutting of stem that is left on the fruit. Packing fruits with stems attached may result in latex stains when stems break off during handling. Fresh latex oozes out of the stem end, thus staining other fruits in a pack. Dried latex is difficult to remove.

Long stems left attached on the fruit should be trimmed prior to packing.

Delatexing/Desapping

Delatexing, also known as desapping, is the removal of fresh latex from the fruit. The steps involved are outlined below:

- Remove bagging material from the fruit. One advantage of bagging is that bags can provide immediate protection from latex staining during harvesting.
- Trim stem neatly close to the base of the fruit using sharp and clean pruning shears.
- Delatexing can be done by either of the following:
    - Inverting freshly de-stemmed fruits on plastic or steel mesh racks, allowing the latex to drip for 30 minutes. Do not invert fruits on burlap sacks as this may lead to coagulation of latex at the base of the fruit.

- Dipping freshly de-stemmed fruits in 1 percent alum solution (one-half kg powdered alum per 50 liters of water) for one minute. Alum enhances coagulation of the latex when fruit are dipped. Plastic crates can be used to hold the fruits during dipping. Allow fruits to dry before packing.

(a) Delatexing of freshly trimmed mango using delatexing tray and (b) by dipping in 1 percent alum solution.

Sorting/Grading

Sorting is the grouping of mangoes based on the criteria of the one classifying and there is no definite set of standards followed. This is the most common practice in selling mangoes. After harvest, mangoes are arbitrarily classified as "Class A" (good quality; for export or for institutional buyers like supermarkets and hotels) or "Class B" (local grade or for domestic market. Grading on the other hand, refers to the classification of mangoes based on the standard criteria accepted by the industry.

Mangoes that are of good quality and safe for consumption are mature, clean, well-formed, free from insect and disease damage, free from mechanical damage such as cuts, abrasion and punctures, free from microbial, chemical and physical contamination. Defects, if any, are only minimal as in the case of windscar and latex stain.

(a) Good and (b) poor quality mango fruits.

Mango fruits with the following defects, on the other hand, are most likely to be rejected:

- Insect damaged due to cecid fly, fruitfly and thrips.
- With pre-harvest damage such as scab and sooty mold, harvesting and handling damage such as latex stains and burns, bruising, abrasion, compression, cuts and punctures, and prematurely ripened.

Classifiers or sorters should practice good personal hygiene such as washing their hands with soap or detergent before handling the produce. There should also be provision for workers' comfort during sorting/grading since uncomfortable sitting positions during sorting may lead to fatigue hence inappropriate sorting procedures.

Inappropriate sorting and packing conditions.

Postharvest Disease Control

Anthracnose and stem end rot are the two most important post-harvest diseases of mango fruits. These two diseases cannot be detected at the green stage and symptoms of infection appear only as the fruits ripen. An integrated pre- and post-harvest disease management protocol will ensure adequate control of these diseases.

Common post-harvest diseases of mango: Anthracnose, stem end rot and fruit rot.

Hot water treatment (HWT) is by far the most effective post-harvest treatment against the two diseases mentioned earlier. HWT is a non-chemical method of controlling decay. Additional benefit gained from HWT is the removal of fresh latex and dirt adhering on the surface of mangoes.

During HWT, green mature fruits are dipped for 5 to 10 minutes in water heated to 52–55 °C. At this temperature range, the disease-causing organisms are killed without injuring the fruit. The effectiveness of HWT is reduced if the temperature drops below 52 °C and while temperature above 55 °C will cause scalding of the peel of the dipped fruits.

Reduced incidence of decay with hot water treatment of mango at the green stage.

Procedure for HWT

- Fill the tank with enough water to completely immerse the crate with fruits in heated water (about 3/4 of the total depth of the tank). Use only clean water for HWT.
- Heat the water to 55 °C. Use the thermometer to check the water temperature:
    - Temperature should be checked at different points of the tank.
    - If temperature is uneven, check if the water pump is functioning properly and water outlets are not blocked.
- Put the mangoes to be treated in plastic crates with holes on all sides and at the bottom. Do NOT overfill the crate. The crate protects the fruits from heat injury resulting from contact with the hot sides and bottom of the tank.
- Dip the crates in hot water for 5 or 10 minutes:
    - Stir the water occasionally (if there is no pump to circulate the water) to have uniform temperature within the tank.
    - The treatment can be shortened to 5 minutes if the mangoes are produced during the dry hot period when conditions are not favorable for infection in the field.

- Remove the crates from the tank:
    - Cool treated mangoes for 10 minutes in clean tap water if fruits are to be brought to distant markets. Cooling is needed since HWT hastens ripening of the fruit.
    - If fast ripening is desired, then fast cooling is not necessary.
- Allow fruits to cool and dry before packing. Faster drying can be achieved by placing crates of mangoes in front of blowers.

Hot water treatment of mango.

Important Reminders during HWT

- Fruits should be treated within 36 hours after harvest to get the maximum benefit of disease control.
- Hot water treatment is most effective against stem end rot when applied while fruit is still fully green.
- Do not allow latex to dry on the fruit before applying HWT since this will aggravate latex burns that develop later.
- Change the water if it becomes contaminated with latex and dirt. Dipping fruit in dirty and latex-contaminated water can increase the incidence of fruit and lenticel damage that develop during ripening.

Packaging and Transport

Packaging

Proper packaging is essential in maintaining produce quality during transport and subsequent handling. The basic functions of packaging are shown in figure.

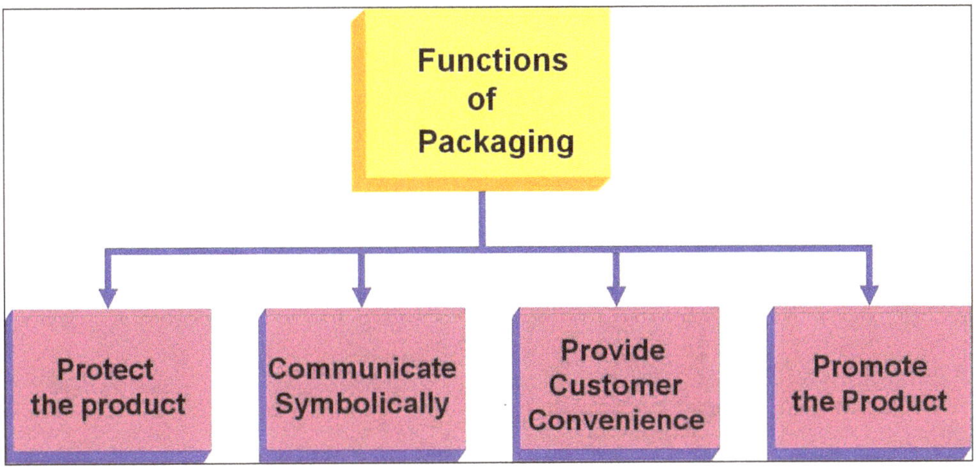

Basic functions of a package.

Packaging materials should provide adequate protection to its contents, should facilitate convenience in handling the produce, attract consumers and sell the contents, and inform the consumers about the produce inside the pack.

Packaging Materials for Mangoes

Rigid containers such as plastic crates are highly recommended for fresh mangoes since they provide adequate protection against compression damage. They have smooth inside finish and can be easily cleaned. They are also stackable and reusable/returnable. Although more expensive than traditional packaging containers, plastic crates with long-term use (5-6 years), the packaging cost per kg of produce is relatively cheaper.

Plastic crates for mango.

Semi-rigid containers such as bamboo baskets are also used but these should NOT be over-packed. Overpacking results in compression damage. Baskets should be lined with clean paper, jute sack or any suitable cushioning material to prevent abrasion and punctures on the fruit since these containers usually have a rough inside finish.

Cartons or fiberboard boxes should be equipped with a vertical divider in the middle and with ventilation holes at the sides to prevent heat buildup inside the carton. The physical strength of fiberboards decreases as they absorb moisture with time. Recycled cartons should NOT be over packed and over stacked as this result in compression damage and the incidence of premature ripening.

Cartons or fiberboard boxes for transporting mangoes to local markets: (a) Overpacked and recycled carton with no ventilation; (b) Cartons containing 18–20 kg mangoes.

While awaiting transport, filled cartons should not be used as a "resting" area as this will lead to damage particularly compression and bruising which may not be apparent at the green stage but will show up later in the handling chain especially when fruits are already ripe.

Considerations in the use of Plastic Crates

- Hygiene: Plastic crates should be thoroughly cleaned with soap/detergent after use.
- Handling: Handle with care during loading, stacking and unloading; do not drop. Don't use as seats when sorting.
- Storage: Store in a clean area that will prevent harborage of insects and rodents. Store separately from chemicals and farm machinery to prevent contamination. Crates should not be left exposed to the external environment since they will readily wear out.
- Do not use as storage containers for chemicals (fertilizers and pesticides) if used for fresh produce.

Transport

The main objective of transportation is to ensure that mangoes will arrive in good condition at the final market. Mangoes must be transported between several points in the supply chain:

- Field to collection center.

- Collection center to wholesale market.
- Wholesale market to retail market.

At all of these stages, good transport practices are described below must be followed:

- Handle containers gently; they should not be dropped or thrown on to each other.
- Containers at the bottom of the stack, should not be used as steps to allow stacking to a greater height especially if semi-rigid containers like cartons are used.
- Allow air to circulate in the stacks or piles of packaged produce by providing space in between stacks. If canvass is used as cover, provide space for air to pass through at the bottom and top of stack; Use light colored material as cover as this will reflect heat.

Cartons/boxes used as steps to load packages in a vehicle.

- Minimize delays or facilitate the transfer of packages from one part of the market to another; four-wheeled hand trolleys will minimize damage during unloading and transfer of produce packed in plastic crates.

Four-wheeled trolley facilitates transport of produce.

- Observe cleanliness of the transport vehicle; produce safety is compromised when:
    - There are decaying remains of produce from the previous shipment.
    - Insects and rodents nesting in the vehicles.
    - Vehicle used as storage area of farm implements when not in use.

Transport vehicle used as storage area of farm implements and tools.

Handling at Wholesale and Retail Markets

The wholesale and retail markets serve as the outlets of mango farmers, collectors and other traders. The basic rules that should be observed are as follows:

- Unload containers from the transport vehicle under cover/shade with careful handling to minimize mechanical damage.
- Re-sort mangoes using sorting table.

Mango fruits in retail display and during temporary holding in the wholesale market.

- Discard culls properly.

- Re-grade mangoes according to size, appearance and stage of ripeness as the case maybe, depending on the requirement of the target market.
- Display mangoes on clean retail shelves or in clean containers.
- When mangoes cannot be sold in one day, keep unsold mangoes in storage with proper ventilation.

POSTHARVEST HANDLING OF BERRIES

Strawberries, raspberries, and blackberries are highly appreciated fruits due to their unique taste and high content in antioxidant and bioactive compounds. They are rich in phenolic compounds, mostly flavonoids and anthocyanins, which are responsible for fruit color and can exert antioxidant, antimicrobial, anti-inflammatory, anticancer, and cardio protective effects. However, berries have a short storage life, as a result of their high respiration and softening rate, and susceptibility to mechanical damages and decay. As berries are considered non-climacteric fruit, they must be harvested at, or near to full maturity, because they will not continue to ripen normally once detached. At this stage, the fruit presents appropriate organoleptic attributes but may become softer and more sensitive to mechanical damage. Thus, it is crucial to be extremely careful during harvest and postharvest handling and to sort, grade, and pack the berries in the field, avoiding excessive manipulation of the fruit. The most extended methods to maintain quality during the postharvest period are prompt precooling and storage at low temperatures. Modified and controlled atmospheres with up to 20-kPa CO_2 and 5–10-kPa O_2 reduce microbial growth and delay senescence but can affect bioactive compounds with a cultivar-dependent response observed for these technologies.

Berry fruits include, among others, strawberries (Fragaria ananassa), raspberries (Rubus idaeus), and blackberries (Rubus spp.). These fruits are characterized by their acidic taste and can be consumed fresh or frozen. Fresh fruits are mainly consumed locally and are available only in the ripening season, except countries from South America, like Colombia or Ecuador, where the production occurs all year round. Berries are also available as processed products like refrigerated fruit pulp, jams, juices, and nectars. What's more, due to their high content in antioxidant and bioactive compounds, they can be considered as functional foods. In effect, different studies conducted on berry fruits report antioxidant, antimicrobial, anti-inflammatory, anticancer, and cardioprotective effects, which were attributed to their high content in bioactive compounds, mainly different phenolic compounds.

However, the manipulation of these fruits presents a series of challenges: berries lack a protective peel and are highly perishable, mainly because of their susceptibility to mechanical damage, water loss, and fungal decay. What's more, berries are considered non-climacteric fruit, which implies that they need to be harvested at, or near to, full maturity as most of them will not continue to ripen normally once detached, and eating quality will not improve after harvest. In some cases, they can color in storage but if they are harvested too early, texture, sweetness, and acidity fail to fully develop.

Fruit quality for the market is largely determined by physicochemical parameters like size, full color, gloss, firm and crisp texture, absence of decay, injuries and bruises, a balance between sweetness and acidity, green sepals, and typical aroma. At the same time, the main causes of loss and rejects include weight loss, presence of bruises and cuts, symptoms of mold and decay, color changes, juice leakage, and sepal wilt.

To get the maximum quality at harvest and maintain this quality during transport and commercialization until the fruit is consumed, it is essential to harvest berries at the optimum stage of maturity. In this sense, the UNECE Standard FFV-57 concerning the marketing and commercial quality control of berry fruit establishes that "Berry fruits must be sufficiently developed and display satisfactory ripeness according to the species but must not be overripe," emphasizing the need to harvest at the appropriate maturity stage for each type of fruit.

Harvest

In order to avoid excessive manipulation and damage to the fruit, berries for the fresh market should be hand-harvested, sorted, graded, and packed in the field, directly into the final container. Fruit ripeness at harvest and fruit handling are two critical factors in the postharvest keeping quality. In fact, the stage of maturity at harvest largely affects the shelf-life of berries, their storage behavior, and sales probability. Immature fruit may have a longer storage capability but are unlikely to develop appropriate organoleptic characteristics while shelf-life of over-mature fruit is generally very short as the susceptibility to decay also increases.

Fruit of blackberries showing different maturity stages.

As berries ripen quickly but non-uniformly, it is crucial to harvest frequently (daily, or every 2–3 days, depending on weather conditions and area of production) and also train pickers to identify the proper ripening stage and in the correct harvest practices to avoid damages to the fruit.

Ideally, the fruit should be harvested early in the morning, after the dew is off the berries or in the evening when the temperatures are cooler. Berries should not be touched before harvest because they are extremely fragile and easily damaged during harvest, for example, by finger pressure. Only sound berries with good appearance should be placed in the packages, and once harvested, fruit must be protected from exposition to direct sunlight. Rotten fruit must be picked off plants and discarded far from the marketable berries to avoid contamination of the latter while small and overripe fruits can be used for processing.

Harvest Maturity

Different maturity indexes can be used for determining berries' optimum harvest date. However, harvest maturity is mainly determined by fruit surface color and most standards require for strawberries that more than one-half to three-fourth of the surface to be colored. In the case of raspberries and blackberries, the fruit must present a completely red and a bright, dark purple/black color, respectively. Color is also the main criterion used by the consumer to judge fruit quality. Besides color, blackberries, and raspberries should pull easily from the receptacle yet being still firm. Regardless of the berry considered and in addition to color, appearance (size, shape, and absence of defects), firmness, flavor (soluble solids, titratable acidity, and flavor volatiles), and nutritional value (vitamin C) are all important quality characteristics that must be considered.

Several studies have shown that the color of these berries can change during storage even if the fruit are harvested at early stages of color development. However, the changes in sugar and acid content of these unripe fruits are not enough to make them suitable for fresh consumption. On the other hand, suitability for selling raspberries declined rapidly with increased ripening stage, and thus, the fruit should not be stored and must be sold and consumed immediately after picking.

In any case, it is recommended to avoid mixing different ripening stages in the same pack, as this practice is usually rejected by consumers at the marketplace. At an industrial level, fruit selection is based both on external attributes such as intense red color and color distribution, fruit size and shape, and absence of physiological defects and on internal quality parameters including sweetness, acidity, and flavor. For an acceptable flavor, a minimum of 7% soluble solids and/or a maximum of 0.8% titratable acidity are highly recommended for strawberries. Similarly, the Ecuadorian Quality Standard NTE-2427 for Andean blackberries (Rubus glaucus Benth) establishes a minimum of 9% soluble solids, a maximum of 1.8% titratable acidity and a minimum of 5 for the maturity index (total soluble solids/titratable acidity).

Packing

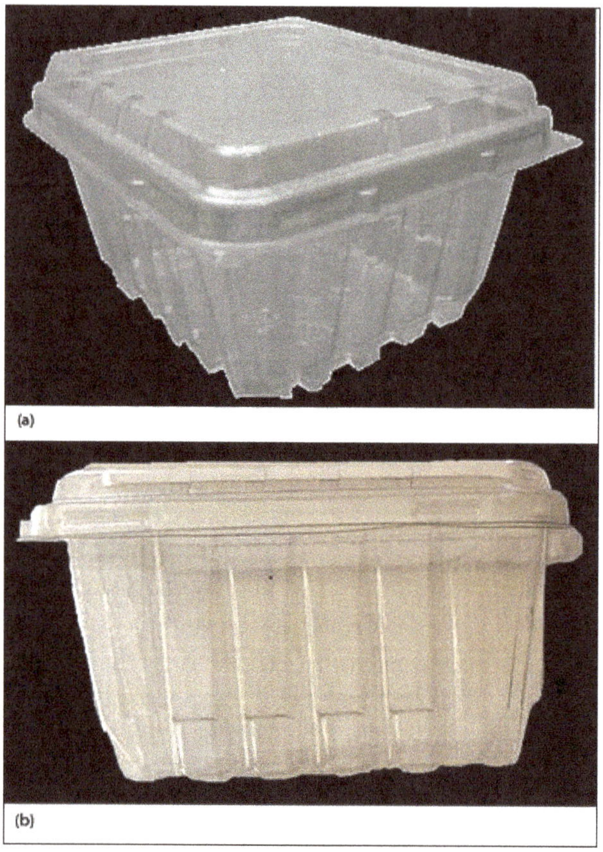

(a) Plastic clamshells of 250 and (b) 500 g for the packing of raspberries and blackberries, and strawberries, respectively.

The containers most commonly used at the supermarket for raspberries and blackberries are plastic clamshells containing 250 g of fruit. In the case of strawberries also, containers for 500, 1000, and even 2000 g of fruit are used, depending on fruit size. Pulp and wooden containers are also used, but they present the disadvantage that stain easily, and wooden containers are also expensive. Regardless of the material, wide and shallow containers are preferred to deep containers and no more than three layers of fruit should be included in each package as the fruit in the bottom may be crushed by the fruit on top.

Plastic clamshells present some advantages:

- They are solid and thus give protection to the fruit from mechanical damage, they do not stain.
- They are inexpensive.
- They are usually clear or transparent, they allow consumers to inspect all the fruit at the time of purchasing.

Baskets used for blackberries harvest.

The containers must be vented on top and sides and have a lid to reduce mechanical damage and moisture loss. On the other hand, the main disadvantage of these packages is plastic disposal after use. In some countries, it is still also a very common practice the use of baskets or buckets, sometimes containing up to 12–15 kg of fruit. Customers pay for the first basket and bring them back in following purchases. This kind of containers are not appropriate as they are usually not washed or disinfected before reuse and thus lack hygiene and can accumulate fungal spores. Furthermore, the excessive weight of fruit causes damage to the fruit located at the bottom which usually collapses. In order to absorb the juice leaked from the fruit, non-food grade, periodic paper is frequently added at the bottom of the baskets, underneath the fruit, worsening the situation. Some efforts are being made to eliminate these kinds of practices and to replace these containers by cartons containing plastic clamshells or cardboard boxes.

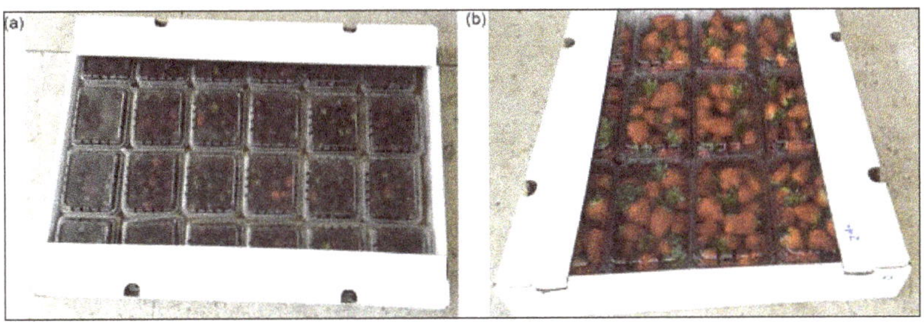

(a) Cartons containing plastic clamshells of 250 and (b) 500 g, for raspberries/blackberries and strawberries, respectively.

Precooling

Precooling, consisting in rapid removal of field heat immediately after harvest, is essential to maintain fruit quality and control decay. For example, strawberries rapidly cooled down to 0 °C showed threefold the storage life of those fruit maintained at 10 °C. Field heat is often removed using forced air cooling, where rapidly moving cold air is forced through pallets of fruit to lower fruit temperature to 0 to 1 °C within 2 h of picking. This method is preferred to room cooling, as forced air can cool berries to 1 °C within an hour, whereas room cooling may take up to 9 h.

High relative humidity (85–95%) should be maintained within the refrigerated rooms, but free moisture on the berries or in the containers must be kept to a minimum as, to reduce fruit rot, the berries must be kept dry. Precooling conditions for blackberries are forced air cooling to 5 °C within 4 h and fruit should be transported at refrigeration temperatures of 5 °C or less. Raspberries should be forced air cooled to 1 °C, no later than 12 h after harvest.

Storage

Strawberries, raspberries, and blackberries are highly perishable due to their relatively high water content, high physiological postharvest activity, and susceptibility to fruit rot and darkening. The high respiration rates of these fruits cause changes in texture, color, flavor, and nutritional content during storage, and such changes are crucial for the determination of fruit quality and consumer's acceptability. Their short storage life is also the result of decay caused by rot-causing pathogens and quick softening rates. Botrytis cinerea, a necrotrophic fungus that causes gray mold rot, is one of the main pathogens responsible for postharvest decay in berries. The symptoms of disease are soft rot with a collapse and water soaking of parenchyma tissues, followed by the appearance of gray masses of conidia.

Botrytis cinerea growth on stored blackberries.

The presence of ethylene in storage can stimulate respiration rates and gray mold growth. Moreover, color of raspberries can be adversely affected by ethylene as it causes darkening of the red fruit to purple-red.

Another physiological disorder that can affect berries during storage is water loss, which in turn causes fruit shriveling, loss of gloss, and plays an important role in anthocyanin degradation. Water loss accelerates senescence of the fruit and the maximum permissible amount of water that can be lost (based on weight loss) from the fruit before becoming unmarketable is 6%. During postharvest handling of the berries, water loss can be reduced by prompt precooling and adequate packaging and storage at optimum temperature and relative humidity.

At present, the most extended methods used to maintain quality and bioactive compounds stability and to control decay of fruits and vegetables are postharvest washing, rapid cooling immediately after harvest, and storage at low temperatures. Furthermore, postharvest diseases are normally controlled by the use of synthetic fungicides and storing under controlled or modified atmospheres with high CO_2 concentrations. However, these methods present some limitations. Berries washing before retail is not recommended because the skin of the fruit may be damaged easily and the drying period delays precooling and enhances infections by pathogenic microorganisms.

Likewise, chemical fungicides may exert several negative effects on food safety and the environment, and there are public concerns about environmental pollution, possible contamination of berries by fungicide residues, and the inability to control fungal diseases because of the appearance of fungicide-tolerant strains of pathogens.

Temperature

One of the main factors affecting the storage shelf-life and quality of fruit and vegetables is temperature, as it regulates the rate of all the metabolic processes that occur in these products. Low temperatures slow down fungal growth and, at the same time, reduce respiration rate and water loss and, therefore, delay ripening and senescence processes. As these berries are insensitive to chilling injury, extending the shelf-life of berry fruit is often achieved through low temperature with optimum storage conditions for strawberries, raspberries, and blackberries being 0 °C and 90–95% relative humidity.

Storage temperature is one of the key factors in suppressing fungal decay and influencing the stability of phenolic antioxidants in fruits during postharvest storage. Also, temperature management is the most important factor to be taken into account to retain the initial ascorbic acid content during storage. Nevertheless, even when temperatures around 0 °C are considered the best for berries' storage, the distribution in trucks and shops, the commercialization, and the storage in consumer households usually occur at higher temperatures, which can affect the berries shelf-life and their physicochemical quality and nutritional value, in terms of soluble sugars, vitamin C, and antioxidant compounds.

Despite the already-known positive effects of low temperatures on postharvest shelf-life and quality of fresh fruits and vegetables, contradictory results can be found. Cooling of fruit at 0 °C can be detrimental for short-term sales, as fruit appearance may be duller, and condensation of fruit during re-warming might result in greater decay incidence. Jin et al. indicated that strawberries stored at 10 °C had higher antioxidant enzymes activities, higher level of phenolics and anthocyanins contents, and stronger oxygen radical scavenging capacities than those stored at 0 or 5 °C, and Kalt et al. found that low temperatures could affect anthocyanin synthesis during storage of small fruits. Similarly, anthocyanin and ascorbic acid biosynthesis was delayed in three strawberry cultivars stored at 6 °C in comparison with storage at 16 °C, while the contents of flavonols, ellagic acid, and total phenolics were not affected by the temperature lowering.

In blackberries a reduction in total anthocyanin content (TAC) after 18 days at 3 °C, while Wu et al. did not see a clear tendency in the evolution of anthocyanins during 7 days at 2 °C. On the contrary, in Andean blackberries harvested at the light and dark-red maturity stages, we observed an increase in total anthocyanin content during 10 days of storage at 8 ± 1 °C and similar results were reported who observed that TAC increased after 15 days of storage at 1 or after 13 days at 1 plus 2 days at 20 °C.

On the marketable quality of strawberries was preserved at 4 °C for a prolonged period of time in comparison with storage at room temperature, while higher antioxidant capacity values were maintained at the lower temperatures, as opposed to 25 °C. Similarly, storage of strawberries at 1 °C together with moisture loss control reduced losses of total ascorbic acid by 7.5-fold compared to fruit stored at 20 °C. These authors concluded that even short periods at ambient temperature without control of water loss could result in considerable losses of total AA in strawberries. Moreover, Shin et al. reported that the best temperature for visual appearance of strawberries was 0.5 °C, but for short-term storage periods (up to 4 days), it was also possible to use moderate temperatures of 10 °C. This temperature was useful to delay fruit ripening compared to room temperature and, at the same time, provided a balance between sensory attributes and those associated with the nutritional status of the fruit.

In Andean blackberries, we also observed that by storing the fruit in refrigerated storage, weight and firmness loss were reduced and microbial growth was delayed in comparison with storage at room temperature. What's more, the refrigerated fruit presented higher scores in the sensory analysis and, the total phenolic content and the antioxidant activity of the fruit were not affected by the cold storage.

Modified and Controlled Atmospheres

Modified atmosphere (MA) and controlled atmosphere (CA) refer to any atmosphere different from the normal air and usually involve atmospheres with reduced O_2 and/or elevated CO_2 levels. The difference between them is that CA is strictly controlled during all time.

Both MA and CA can be used for the storage, transport, and packaging of different types of food in compliment to low temperatures to extend their shelf-lives after harvest. Exposure of fresh horticultural crops to low O_2 and/or elevated CO_2 atmospheres within the range tolerated by each commodity reduces their respiration and ethylene production rates and therefore results in several beneficial effects such as delay of ripening and senescence and associated biochemical and physiological changes, reduction of sensitivity to ethylene action, alleviation of certain physiological disorders such as chilling injury, direct, and indirect control of pathogens, and consequently decay incidence and severity. On the contrary, if horticultural products are exposed to O_2 concentrations below, and/or CO_2 concentrations above their optimum tolerable range, the initiation and/or aggravation of certain physiological disorders, irregular ripening, increased susceptibility to decay, development of off-flavors, and eventually the loss of the product can occur.

Modified atmospheres (MA) and controlled atmospheres (CA) with elevated (15–20%) carbon dioxide and 5–10% oxygen concentrations reduce the growth of Botrytis cinerea (gray mold rot) and other decay-causing organisms. In addition, it reduces the respiration and softening rates of berries, thereby extending postharvest life. Nevertheless, further reductions of O_2 concentrations to 2 kPa had no benefit and may cause fermentation of the fruit.

In addition, off odors can be produced if the fruits are kept under high CO_2 atmospheres for more than 4 days as a result of anaerobic respiration and the effect on the flavor preservation of these fruit is not clear. Several authors reported changes in pH, titratable acidity, total soluble solids, sugars and organic acids, and fermentative metabolites after storage under CO_2-enriched atmospheres. In effect, different fermentative volatiles (acetaldehyde, ethanol, and ethyl acetate) were found after storage of strawberries in air + 20-kPa CO_2 at 2.8 °C. Among aroma compounds, esters are apparently the volatiles most affected by CO_2-enriched atmospheres.

Anthocyanin synthesis continues after harvest, but it is inhibited in fruits stored in high CO_2 concentrations. According to Holcroft and Kader, high CO_2 concentrations together with low O_2 concentrations can also affect adversely total ascorbic acid and anthocyanin contents and, thus, have a negative impact on fruit color and nutritional value. Conversely, the firmness, the external color, and the total phenolic compounds content of Selva strawberries were not affected by storage atmospheres with up to 20-kPa CO_2. In both, strawberries and raspberries, a cultivar-dependent response to changes in the storage atmosphere was observed.

An atmosphere of 12.5-kPa CO_2 and 7.5-kPa O_2 was effective in reducing decay in red raspberries and elevated concentrations of CO_2 together with reduced concentrations of O_2 were shown to inhibit mycelial growth and spore germination of B. cinerea and other fungi responsible for postharvest decay of fruit. What's more, raspberries stored in 10/15-kPa O_2/CO_2 presented a more attractive color in comparison with fruit stored in air. The use of high and medium barrier materials delayed senescence and did not affect negatively the nutritional and antioxidant properties of red raspberries stored at 4 °C. However, fermentative volatiles were found for these fruits, especially when an oxygen absorber was included in the packages.

Alternative Methods for Decontamination of Strawberries

The need to minimize chemicals use has encouraged the rapid development of alternative techniques. One of the new approaches is the use of 'generally recognized as safe' (GRAS) products, such as UV radiation and ozone, due to minimal concerns about their environmental impact and low residues in the treated commodity.

UV-C Radiation

One strategy that can be an adjunct to refrigeration is the exposure of fruits to hormetic doses of UV-C radiation, a physical treatment that has been tested in strawberries and other fruits to control postharvest diseases and delay some ripening-associated processes. Hormesis has been defined as the use of potentially harmful agents at low doses in order to induce a beneficial stress response.

UV-C seems to have a direct germicide effect on pathogens and an indirect effect by inducing defense mechanisms in the plant tissues. Irradiation with UV-C is known to stimulate the phenylpropanoid pathway in several fruits, mainly by the induction of phenylalanine ammonia lyase (PAL), a key enzyme in this pathway. The compounds synthesized by this pathway are implicated in a protective role against pathogens through reinforcement of plant cell walls, direct inhibition of growth, and/or inactivation of enzymes that contribute to tissue maceration.

Particularly in strawberry fruit, different UV-C doses increased enzyme activity, the antioxidant capacity and total phenolic content during storage, which correlated with lower fruit decay observed in treated fruit. The synthesis and accumulation of phenolic compounds following irradiation with UV-C could also play an additional indirect role in fruit protection acting as natural substrates of polyphenol oxidase (PPO). One of the proposed roles of the reaction products of PPO (quinones) in plant defense is their action as bactericidals and fungicidals. It was found that postharvest UV-C treatment, a few hours prior to inoculation with B. cinerea, reduced the percentage of fruit infection in strawberries during storage. Irradiation of fruit with UV-C increased expression and activity of several enzymes (PAL, peroxidases, PPO, chitinases, and β-1,3-glucanases) which are involved in defense mechanisms against pathogens and abiotic stressors. In another experiment, Figure chemical treatment with UV-C delayed the appearance of gray mold rot in stored strawberries by up to 5 days at both 4 and 13 °C. The treatment with UV-C also enhanced the accumulation of anthocyanins, which in turn contributed to redder and visually more appealing fruit. In contrast, Erkan et al. found little effect of UV-C treatments on anthocyanin content in strawberries.

Use of Ozone

Another emerging technology with potential application in the food industry is the use of ozone as a sanitizer. O_3 can be used for the postharvest treatment of fresh fruits and vegetables, in air or water, or as a continuous or intermittent atmosphere throughout the storage period. Gaseous ozone can be used to sanitize storage rooms and to prevent bacteria, molds, and yeasts development on the food surfaces. It can also eliminate undesirable flavors produced by bacteria and chemically remove ethylene gas to slow down the ripening process.

Ozone gas efficacy to inactivate microorganisms is conditioned by the species considered, its growth stage, the ions present in the air, the O_3 concentration and exposure time, and, the temperature and relative humidity of the room. In air, the reactivity of ozone is greatest with fungi, molds, and some odor-causing chemicals and least with dry spores and bacteria. For optimum efficiency, it is also essential that the gas is thoroughly and evenly distributed quickly. Otherwise, decomposition will occur before the O_3 is able to contact its target.

Washing strawberries with ozonated water (0.3 ppm, 2 min) was an effective treatment to reduce microbial counts and enhance anthocyanin and ascorbic acid retention of these fruits during 13 days of refrigerated storage. Similarly, Zhang et al. reported greater ascorbic acid retention in strawberries treated with gaseous O_3 (4 ppm, 30 min/day) in comparison with untreated fruit. Moreover, strawberries' levels of biothiols were not affected by the treatment with either gaseous-phase or aqueous-phase ozone. Finally, while total phenolic and ellagitanin contents were similar in O_3-treated and untreated strawberry fruit after a storage period of 12 days, the procyanidins and anthocyanins contents were reduced by the exposure to this gas.

In addition to their antimicrobial power, O_3 and UV-C radiation gather other advantages, which turn them into appealing and environmental friendly technologies. Neither ozone nor UV-C leave undesirable residues on food or food-contact surfaces nor create undesirable disinfection by-products. Moreover, the application of these sanitizers in food processing is approved by the code of Food and Drug Administration (FDA) in the USA and is allowed by organic certification.

Results are sometimes contradictory and information about the effects of these decontamination treatments on sensory and nutritional quality or health-promoting composition of treated products is scarce. It should be taken into account that while high doses of oxidizing agents may result in depletion of natural antioxidants, moderate or low doses of oxidative stress were shown to cause a protective response, enhancing the level of endogenous antioxidants.

Transport

Berries must be transported in clean and well-maintained trucks, and it is crucial to maintain the fruit cold and wrapped during loading, unloading, and transportation. In order to ensure a proper circulation of the cold air, the flats or boxes must be stacked on pallets and without touching the truck walls. Frequently, rural roads are not in optimum conditions causing bruises and abrasion due to the truck vibration and the impacts between the packed fruit and between the fruit and the walls of the packs. These mechanical damages can be minimized by stabilizing the load on every pallet, for example, by using stretch film and by using trucks with air suspension systems.

When refrigerated transport is used, it must be considered that trucks' mechanical refrigeration equipment is designed to maintain temperature but they do not have the capacity to lower the temperature of the produce. So, it is very important to achieve the proper cooling of the product before loading. Finally, to avoid condensation on the fruits, berries should be allowed to warm only when they are ready for display to consumers and before removing the plastic wrap over the flats.

POSTHARVEST HANDLING OF TOMATO

Tomato (Lycopersicon esculentum Mill.) is a vegetable crop popularly consumed. It is commonly used for table consumption in fresh form or as an ingredient in many cooked dishes. Aside from its economic value, tomato is good for human nutrition since it is a source of Vitamins C, A and K, potassium, and carotenoids such as lycopene and carotene which act as antioxidants.

Importance of Postharvest Handling

Nature of losses in tomato: Decay, over-ripening and mechanical damage.

Proper post-harvest handling is important in maintaining the quality and insuring safety of the fruit while being brought to consumers on time, and in meeting buyers' specifications and trade requirements. Post-harvest losses – Losses in quantity and quality occur between harvest and consumption. In tomatoes, losses occur due to immaturity, over-ripening, mechanical damage, and decay. These losses can be attributed to poor harvesting method, rough handling, improper packaging and poor transport conditions. Unless these losses are minimized, the gains from production and the potential income cannot be realized. Post-harvest losses represent a waste of the resources – land, labour, energy, water, fertilizer, etc. that went into producing the crop. Every effort must, therefore, be made to minimize these losses.

Consumers are now increasingly paying attention to quality. They are looking for and are willing to pay a premium price for good quality produce that is safe. Preservation of the nutritive value goes hand in hand with the prevention of quality deterioration. With changing consumer tastes and lifestyles and the continuous growth of supermarkets and the demands from institutional buyers, more attention in post-harvest handling of tomato to satisfy their demand for better quality and safe produce.

Good quality tomatoes are required by institutional buyers like supermarkets.

Characteristics of Tomato Fruit affecting Postharvest Life

In general, post-harvest technology can only maintain, not improve or little if any, the quality of harvested fruit. Therefore, the main objective of any post-harvest technology is to keep fruit quality and safety as high as possible until it reaches the final consumer.

Tomato Fruits Undergo Programmed Changes

Programed change in tomato like ripening.

The tomato fruit, being a living tissue is subject to continuous change until it completely deteriorates. The process of ageing and eventual dying in living tissues is called senescence. There are programmed changes inside the fruit that influence appearance, flavor, texture and nutritive value that cause them to age. While most changes are desirable like those occurring during ripening, some changes are deteriorative. These post-harvest changes cannot be stopped but can be slowed down within certain limits.

Tomato Fruits Lose Water

Tomato fruits consist mainly of water and when harvested, it can no longer replace the water that is lost. Conditions like high temperature and low relative humidity result in water loss hence the loss in marketable weight. Water loss from the fruit results also in shriveling. Exposing fruit to the sun results in rapid water loss.

Tomato Fruits are Prone to Injury and Decay

Improper handling like packing in bamboo baskets and sitting on packed tomatoes during transport lead to damage on the fruits.

Tomatoes are also prone to injury. When the fruit is damaged, the biological processes like respiration and ethylene production proceed at a very fast rate leading to rapid deterioration. Practices like packing in bamboo baskets with rough and pointed edges and siting on packaged tomatoes during transport will injure tomatoes. The damage may not be seen at the green stage but may show up later during retail as compression and discoloration in the flesh.

Tomatoes are also susceptible to attack of insects and decay-causing organisms which can eventually result in faster deterioration. Rough handling may create wounds which serve as entry points for decay-causing organisms. Wounds are also avenues for water loss and can speed up the ripening process.

Harvest Management Practices

Harvesting at the proper stage of maturity, the manner in which the fruit is detached from the tree and the time of harvesting are important considerations during harvesting.

Harvest Maturity

External and internal appearance of mature green tomatoes.

Tomatoes can be harvested at any stage of maturity and the distance from the production site to the consumption area should be considered. Tomatoes are ready for harvest as soon as they are mature but still green. Representative samples of green mature fruits are cut crosswise and if the seeds slide without being cut, then the tomatoes of the same variety can be considered mature. Immature fruits when picked, fail to develop full color and flavor and deteriorates easily.

Time of Harvesting

Pick tomatoes during the cooler part of the day. It is recommended to harvest up to noontime and to keep the harvested fruits in a shaded area soon after picking. A rise in fruit temperature occurs when the fruits are left exposed to the sun. Heating up will result in accelerated ripening of fruits.

Exposure of newly harvested tomatoes will result in heat build-up and accelerated ripening.

Harvesting Method

Harvest tomato carefully by hand picking. Mature fruit is best removed by a "lift, twist and pull" technique. Harvesters should have trimmed fingernails to prevent injury to the fruit. The use of clean gloves during picking will minimize damage on the fruit caused by fingernails. Harvesters should observe good personal hygiene during harvesting. Harvested tomatoes should be placed in clean collecting containers such as plastic pails with smooth inside finish and then transferred to field containers like the stackable plastic crates once filled.

Hand-picking of tomato fruit and use of gloves during harvesting prevents damage on the fruit.

The fruit should be handled carefully during harvesting. Practices like throwing tomatoes into the container, dropping and dragging of containers during hauling causing fruits to bump against each other must be prevented. These poor handling practices will result in both visible and non-visible damages like internal bruising. Visible injuries can be in the form of cuts, punctures or abrasion. Internal damage will appear later as brown or black discoloration in the seed area or manifested as faster ripening and increased susceptibility to decay.

Collecting pails and stackable plastic crates are appropriate field containers of harvested tomatoes.

Post-harvest Handling Operations

Post-harvest operations refer to activities done to the fresh produce in preparation for marketing to meet the requirements of the target market. These operations can be done in the field, in collection centers or in a packinghouse. The packing area should provide adequate protection from sun and rain, kept clean at all times, and pet animals should be kept away from the packing area. All workers should maintain a high degree of personal hygiene and where appropriate, should wear suitable protective clothing and head covering.

Cleaning/Washing

Newly harvested dirty tomatoes should be washed in chlorinated water.

In places where tomatoes are not staked or trellised and grown during the rainy season, soil particles may adhere on the fruit. Soil-borne decay-causing microorganisms can cause fruit rots thus the need to clean them prior to marketing. A disinfectant like sodium hypochlorite (ordinary bleaching solution) can be added to the clean water (4 tablespoon per gallon of water). Washing in 2 percent sodium bicarbonate (20 grams baking soda per liter of water) was shown to reduce storage rot. Fruits should be dried properly prior to packaging. Tomatoes can also be cleaned by wiping them with moist cloth. Insure however, that the water and the cloth used in wiping the fruit are clean to prevent contaminating the fruit.

Wiping of tomatoes with moist clean cloth.

Sorting/Grading

Good quality tomatoes are generally preferred by buyers, thus sorting/grading is a necessary operation. Sorting is the grouping of tomatoes based on the criteria of the one classifying and there is no definite set of standards followed. After harvest, tomatoes are arbitrarily classified as "Class A" (excellent quality) or "Class B" (with minor defects). The names of the grades are determined or specified by the buyer.

Good quality and safe tomato fruits are mature, clean, well-formed, free from insect and disease damage, free from mechanical damage such as cuts, abrasion and punctures, free from microbial, chemical and physical contamination.

Tomato fruits with the following defects, on the other hand, are most likely to be rejected:

- Insect-and disease-damaged.
- Mechanically-damaged such as cuts, punctures, abrasion, compression.
- With preharvest defects such as deformation and cracks.

Sorting is done using a simple sorting table where sorters are positioned on both sides of the sorting table. Sorted tomatoes are packed directly in the container. With the use of sorting table, contact of the fruit on the soil is prevented. Sorting of tomatoes on the ground with plastic sheet as underlay may contaminate the fruit and is inconvenient for the sorter. The sorting tables must be cleaned preferably disinfected with chlorinated water after using.

Common forms of rejection of tomato.

Classifiers or sorters should practice good personal hygiene such as washing their hands with soap or detergent before handling the produce. There should also be provision for workers' comfort during sorting/grading since uncomfortable position during sorting leads to fatigue hence inappropriate sorting procedures.

Sorting of tomatoes using a sorting table (a) which is good practice, and sorting on the ground (b) which may lead to microbial contamination.

Packaging

Proper packaging is essential in maintaining product quality during transport and subsequent handling. The basic functions of packaging are to contain sufficient volume of produce, to protect the contents, to sell the produce, and to inform the buyer about the produce.

Packaging Materials

Rigid containers such as plastic crates are highly recommended for tomatoes since they provide adequate protection against compression damage. They have smooth inside finish and can be easily cleaned. They are also stackable and reusable/returnable.

Although more expensive than the traditional packaging containers, plastic crates have been shown that in long run use (5-6 years), the packaging cost per kg of produce is relatively cheaper.

Plastic bags, plastic sack and red mesh bags do not provide adequate protection to the contents. They are readily damaged and will lead to fruit deterioration. Injuries often incurred include compression manifested as flattened areas and breaks in the skin.

Tomatoes packed in red mesh bags and stacked several layers high resulting in fruit damage.

Considerations in the use of Plastic Crates

Dirty plastic containers as source of contamination; (b) clean plastic crates after use.

- Hygiene: Plastic crates should be thoroughly cleaned with soap/detergent after use. Sanitizer such as sodium hypochlorite reduce microbial load in plastic crate.
- Handling: Handle with care during loading, stacking and unloading; do not drop. Don't use as seats when sorting.
- Storage: Store in a clean area that will prevent harborage of insects and rodents. Store separately from chemicals and farm machinery to prevent contamination. Crates should not be left exposed to the external environment since they will readily wear out.
- Do not use as storage container for chemicals (fertilizers and pesticides) if used for fresh produce.

Transport

The main objective in transport is to insure that tomatoes will arrive in top quality condition to the final market. Tomatoes often undergo three (3) transport stages: (a) Farm to collection center or packing shed, (b) Packing shed to wholesale market, and (c) Wholesale market to retail market. In all of these stages, follow the proper transport practices given below:

- Handle containers gently; they should not be dropped or thrown on to each other.
- Do not seat on top of packed tomatoes during field transport.
- Containers at the bottom should not be used as steps to allow stacking to a greater height especially if semi-rigid containers like cartons are used.
- Do not expose tomatoes to the sun during transport.
- Allow air to circulate in the stacks or piles of produce by providing space in between stacks. If canvass is used as cover, provide space for air to pass through at the bottom and top of stack; Use light colored material as cover as this will reflect heat.
- Minimize delays or facilitate transfer of packages from one part of the market to another; four-wheeled hand trolleys will minimize damage during unloading and transfer of produce packed in plastic crates.

Four-wheeled trolley facilitates transfer of produce.

Observe cleanliness of the transport vehicle; produce safety is compromised when:

- There are decaying remains of produce from the previous shipment.
- Insects and rodents nesting in the vehicles.
- Vehicle used as storage area of farm implements when not in use.

Handling at Wholesale and Retail Markets

The wholesale and retail markets serve as the outlets of tomato farmers, collectors and other traders. The basic rules that should be observed are as follows:

- Unload containers from the transport vehicle under cover/shade with careful handling to minimize mechanical damage.
- Re-sort tomatoes using sorting table.
- Discard culls properly.
- Re-grade tomatoes according to size, appearance and stage of ripeness as the case maybe depending on the requirement of the target market.
- Place tomatoes in a clean storage room in the wholesale market; keeping packed tomatoes near vegetable trimmings, culls and wastes will lead to contamination.

Re-sorting of tomatoes in the wholesale market.

- In the retail market, it is best to sell tomatoes in retail packs since this will prevent frequent handling of customers when they select tomatoes. Frequent handling of customers may also lead to contamination. If retail packaging cannot be done, tomatoes should be placed in appropriate containers during retail display. Displaying tomatoes loose on the ground with plastic mat underlay may not provide adequate protection from contamination.
- When retailing in open-air market and roadside stalls, tomatoes on display should be under shade for sun and rain protection.

Tomatoes in retail packs (in supermarkets) and in retail display in wet markets.

POSTHARVEST HANDLING OF BEAN

Snap bean ranks 6th as the most demanded vegetable. It is produced mostly on small farms. The peak harvest seasons for snap beans are March – April and September – December. As a vegetable, snap bean is high in protein and soluble fiber and low in calories. Snap beans are an excellent source of vitamin K and are a very good source of manganese, vitamin C, dietary fiber, folate, and vitamin B2, copper, vitamin B1, chromium, magnesium, calcium, potassium, phosphorus, choline, vitamin A (in the form of carotenoids), niacin, protein, omega-3 fatty acids, iron, vitamin B6, vitamin E, and contains valuable quantities in absorbable form, of the mineral silicon.

A good quality snap bean pod is fleshy, young and tender, snaps easily, and is free of physical injury and damage caused by diseases and pests.

Importance of Postharvest Handling

Post-harvest handling includes all the primary processes/steps that a harvested crop has to go through to get from the producer to the end consumer. Such processes which include raw material handling, storage, transport, distribution, and marketing, add value to the harvested produce. Post-harvest handling is an element of post-harvest science and technology, the goals of which are to maintain the fresh quality, as well as, assure the safety of crop commodities used as food and to meet buyers' specifications and trade requirements, while in the process, reducing losses.

Post-harvest handling is the weakest link in the supply chain from the producer to the market. Much is lost, both in terms of quantity and quality (including nutritional quality) during postharvest handling, with fruits and vegetables incurring high losses due to their perishable nature. Post-harvest losses in the snap bean industry were reported at over 40 percent, with the marketing sub-sector incurring higher wastage and transaction costs.

Losses in snap bean are the result of the following: poor knowledge about the right harvesting index, thus, a large proportion of the harvested beans are usually over-mature;

poor handling practices – such as the use of plastic sacks for bulk packaging and transportation which results in mechanical damage that serves as entry points for disease-causing organisms leading to rotting of the pods; poor transport practices such as the use of trucks that have no cover thus exposing the produce to direct sunlight and high temperature; absence of low temperature storage facilities and transport systems; and, rough handling practices during distribution in retail markets. Maintaining snap bean quality and reducing quantitative losses is best done through the application of good post-harvest management practices, supported by good post-harvest technologies and through improvement of the post-harvest system.

Characteristics of the Snap Bean

The snap bean, is an immature legume fruit or pod with underdeveloped seeds. Thus it is tender and crisp and snaps easily when hand-broken. If the pod is allowed to mature on the vine, the seeds will develop fully causing the pod to bulge. The mature pod becomes stringy and fibrous indicating poor eating quality, hence, low marketability. As an immature pod, the skin of the snap bean is thin and is very sensitive to even a slight applied physical force resulting in mechanical damage such as bruising and cuts. The damaged portion serves as an entry point for microorganisms that cause rots.

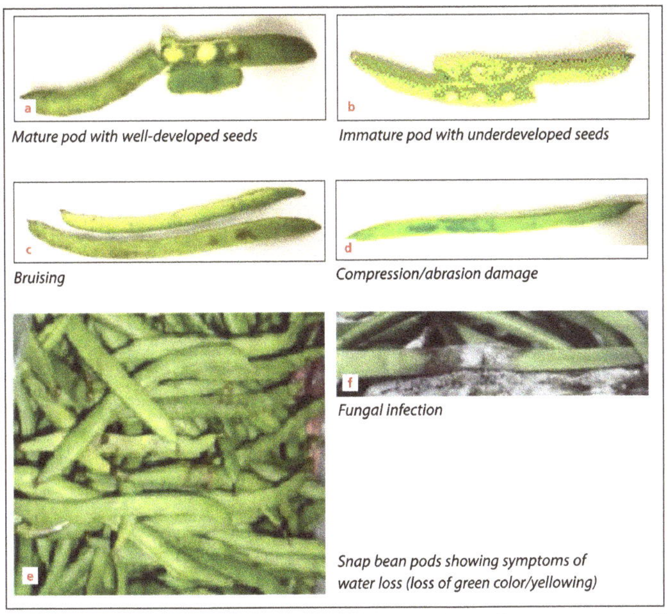

Snap bean pods showing symptoms of water loss (loss of green color/yellowing).

The snap bean contains almost 85 percent water and because of its thin skin, it has the tendency to lose moisture (moisture loss) very quickly. The rate of water loss is highest when the temperature of the outside environment is high, and when the amount of moisture in the air (relative humidity) is low – for example on a hot, dry day-, resulting in loss of crispiness/freshness, softening and consequently yellowing and toughening of the beans.

Due to their high respiration rates and sensitivity to ethylene, snap beans have a short shelf-life and deteriorate rapidly after harvest.

Harvest Management Practices

Harvesting is the process of cutting/removing the pod from the mother plant. There are three important considerations to bear in mind during harvesting:

Harvesting Index

Signs or indications that the snap bean pod has the desirable characteristics intended for its use. The snap bean pod must be harvested when it reaches the desired marketable size and before it becomes too large. In order to ensure that the bean meets consumer demand for quality at harvest, the pod should be green, fleshy, tender, and should easily break or snap when broken apart by the hand. The pod is considered over-mature and of poor quality when the seeds within the pod are fully enlarged and the pod bulges.

Time of Harvest

Time of day appropriate to harvest the pod. Many crops are harvested very early in the morning to take advantage of the cool temperature. For snap beans, the appropriate time to harvest is when the morning dew is off and the plants is thorougly dry.

Harvesting Procedure

Method of harvesting, use of harvesting tool and harvest container – the manner of detachment of the pods and collection of harvested pods. The pods are best harvested manually using a cutting tool (such a pair of scissors or shears). The pod must be held by hand and cut off from the plant. Harvested pods must be directly placed in a field collection container for transporting to a collection point. Plastic crates are one example of a good field collection container.

Snap bean harvested at the right stage of maturity harvested over-mature.

Postharvest Handling Operations

Post-harvest handling operations prepare harvested produce for marketing. Post-harvest handling operations can be done in the field, in collection centres or in a pack-house. It must be well-protected from the sun and rain and kept clean at all times. Animals and pets must be kept away from the harvested beans as they could contaminate the pods with their urine and fecal matter. Similarly, workers who handle the beans must observe good hygiene.

Field Sorting

Culling out produce that is unmarkertable in the field to lower the cost of hauling and to minimize disease contamination. It is recommended that where possible, snap beans are sorted at the farm in order to remove low quality (diseased) or damaged (insect or rat-damaged beans). Sorting in the field would also reduce the cost of hauling the beans to the collection centre.

Snap bean pods are harvested manually using a pair of shears to cut the pods from the mother plant.

Sorting must be done in a shaded area away from the sun. A temporary shed/shelter or the shade of a tree near the harvest area can serve the purpose of providing shade.

A plastic crate is the best harvest container for collection of the pods.

Sorting/Grading

At the packing area, snap beans may be sorted to remove beans with harvest-related defects such as cuts, severe abrasions, etc. before grading. Grading of snap beans is done on the basis of pod size (12 –15 cm in length) or maturity. Sorting and grading can be done at a collection centre or in a packing area equipped with sorting tables and weighing scales. The area in which these operations are done must be clean and well lit. Workers must observe good hygiene and must be able to sit comfortably in order to engage in sorting operations.

Examples of pre-harvest defectsto be removed during field sorting: (a) Beans that are diseased with anthracnose, (b) Insect damaged, (c) Diseased with rust.

Packaging

Bulk packaging of snap beans for transportation to markets is considered the weakest link in the supply chain. Poor bulk packaging of snap beans results in quality defects due to bruising, wounding (cut, puncture, crack, split, breakage), change of shape, compression damage and abrasion. This damage is greater when the packages are either under-filled, over-filled or if the packaging material has rough surfaces, is not rigid, and when the packages are dropped during handling.

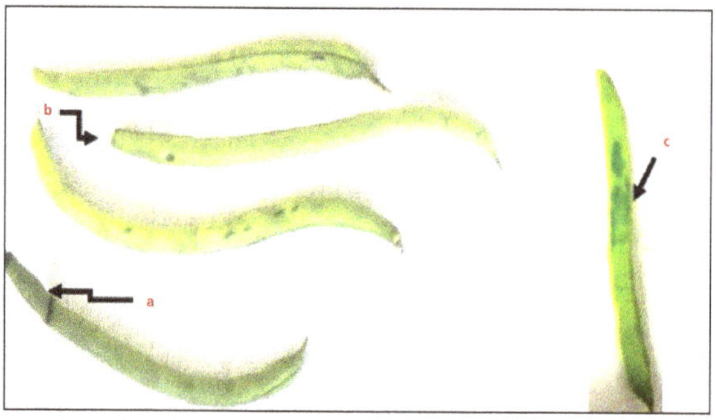

Some harvest-related defects that are sorted out prior to grading: (a) Nail cut, (b) Cut due to excessive (c) Hand-pulling, abrasion damage.

Packaging Materials

Several types, of packaging materials are commercially available for use in handling and transporting fresh produce such as the snap bean. These include wooden crates, stackable plastic crates, polyethylene or polypropylene plastic bags, sacks made from plastic twine, and cardboard box/cartons. With the exception of plastic crates and mesh bags, these packaging materials do not allow for good ventilation, resulting in moisture condensation inside the bag, causing the pods to rot. Nevertheless, these packaging materials are commonly used because of their low cost and ready availability. In most cases, they are not properly used; often, they are overfilled, or are either too large or too small.

Bulk packaging of snap beans makes use of 50-kg mesh bags, or plastic twine sacks.

While the cost of plastic crates is high, they can be repeatedly used over 6 to 10 years. Plastic crates are the best bulk packaging containers for snap beans because of their smooth inside finish, good ventilation that prevents the buildup of heat produced during respiration of the bean pods; and because of the ease with which they can be cleaned. Due to the uniformity of their shape, plastic crates can be stacked securely inside transport vehicles. For best results, beans must be loosely packed in plastic crates in order to allow air flow through the beans inside the package.

When using plastic crates as bulk packaging containers, the following considerations should be borne in mind:

- Hygiene: Plastic crates must be thoroughly cleaned with soap/detergent, rinsed in a sanitizer and dried prior to use.
- Handling: Crates must be handled with care during loading, stacking and unloading; Crates must not be dropped or used as seat during sorting operations.
- Crates must not be used as storage containers for chemicals (fertilizers and pesticides) if used for fresh produce.
- Storage: Crates must be stored in a clean area that will prevent their contamination by insects and rodents. Crates must be stored separately from chemicals and from farm machinery to prevent contamination. Crates should not be left exposed to the external environment since they will readily wear out.

Snap beans are BEST packed in clean and undamaged plastic crates for transportation.

Storage of Snap Beans

Snap beans are sensitive at temperature range of 5–10 °C, hence, they have to be stored at the optimum low temperature condition which is 10–15 °C. At this temperature range, the beans will stay fresh and in good condition for about 2–4 weeks. In the absence of a refrigerated storage system, a simple, low cost evaporative cooler that allows lowering of temperature of the surrounding air by 3–5 degrees, can be used to maintain the freshness of the snap beans for about a week or so. An evaporative cooler works well in hot tropical conditions like that. Three models of evaporative coolers are presented below:

The PHTRC Storage/cooling chamber equipped with an evaporative cooling pad with charcoal as the wetted media. A submersible pump is used to move water from a sump tank to a water distributor at the top of the cooling pad.

Transport

Transport is key in the distribution of fruits and vegetables. Because of their inherent perishable nature, fruits and vegetables must reach destination markets within a short period of time, necessitating rapid and reliable transport systems.

Field-level transportation of snap beans would include the hauling from the farm to the collection center. This is done using a hauling vehicle if the collection centre is far from the farm, otherwise, manual hauling is practiced. Snap beans must subsequently transported from the collection centre to wholesale and to retail markets normally by a three-wheeled cab.

The level of losses that takes place during transport is influenced by the distance between the farm and the market, the number of loading and unloading points, the quality of the road, temperature of the environment, and availability of handling aids.

In order for the snap beans to arrive at their destination in good condition, good transport practices must be observed:

- During loading and unloading, the containers (plastic crates) must be handled gently; they should not be dropped or thrown on to each other.
- Passengers must not stand and/or sit on top of packages during transportation.
- Containers at the bottom of the stack of packages should not be used as steps to allow stacking to a greater height especially if semi-rigid containers like cartons are used.
- Packages of beans must not be exposed to the sun during transport.
- Allow air to circulate in the stacks or piles of bean by loosely packing the crates, and providing space between the stacked crates. If canvass is used as a cover, place the crates on top of a pallet in the vehicle in order to allow for air circulation around the produce. Use light colored material as cover as this will reflect heat.
- Minimize delays or facilitate the transfer of packages from one part of the market to another; four-wheeled hand trolleys will minimize damage during unloading and transfer of produce packed in plastic crates.
- Ensure that the transport vehicle is clean. A clean transport vehicle must:
    - Not contain decaying remains of produce from the previous shipment.
    - Not be infested with Insects and rodents nesting in the vehicles.
    - Not be used as a storage area for farm implements and chemicals.
    - Not be used to transport organic fertilizers.

Handling at Wholesale Markets

Snap beans are usually marketed by traders in wholesale and retail markets. Upon arrival at these distribution channels, certain basic rules should be followed to maintain the freshness and quality of the pods:

- Carefully unload containers from the transport vehicle to the marketing area or under a shaded area.
- Depending on the target market or buyer, re-sort or re-grade the pods and discard any damaged or rotten pods.
- Do not place the packages of beans in dirty areas in the wholesale market to minimize microbial contamination.

Handling in Retail Markets

Snap beans in retail plastic packs with small holes – note the absence of moisture condensation in the bags provided with diffusion holes.

In the retail market, snap beans are best sold in plastic bags. Packaging in plastic bags prevents moisture loss from the pods as well as damage due to frequent handling by customers when they select the beans. A fewpinpricks or small holes on the bags will help prevent the accumulation of moisture in the bags, and will allow the beans to remain fresh for longer periods. Plastic bags must not be over-filled in order to prevent breakage of the pods.

4

Postharvest Storage of Crops

Different crops require different temperatures and techniques to be stored after harvest in order to avoid any degradation in the quality. This chapter discusses in detail the theories and methodologies related to the postharvest storage of different crops such as grains, onions, etc.

Postharvest storage implies a strict control of environmental temperature and humidity in the storing chamber, and it is known that cuticle properties are largely influenced by these two factors. For example, when fruit cuticles from mature tomato fruit were examined for rheological properties, remarkable differences were observed between dry and rehydrated samples. After being dipped in distilled water, extensibility and plasticity of cuticles increased. Furthermore, the rheological properties of cuticles were also sensitive to minute temperature differences in the range from 7 to 30 °C.

The alterations in extensibility were unrelated to cuticle thickness or ultrastructure, and no significant differences in these properties were found between fruit cuticle and fruit skin. These data suggest that tomato fruit cuticles profoundly impact the mechanical attributes of the whole organ, which clearly points out interesting possibilities, such as an influence on postharvest changes in fruit firmness, resistance to mechanical damage, or susceptibility to biotic attack. Analogous results were obtained in a concurrent study in which the mechanical properties of tomato fruit cuticles were investigated for their dependence on temperature and relative humidity in a range of 10–45 °C, and 40% to wet (immersion in aqueous solution), respectively. Isolated cuticles were submitted to uniaxial tension stress, in order to assess several essential mechanical properties including tensile modulus, breaking stress or maximum elongation. These tests showed that stress–strain curves of tomato fruit cuticles were biphasic when humidity values were below wet conditions, but monophasic when cuticles were wet. Whereas maximum elongation was independent of relative humidity and temperature, temperature decreased pure elastic strain and breaking stress. This response was also biphasic, consisting of two temperature-independent phases separated by a transition temperature related to the presence of a secondary phase transition in the cutin matrix of the cuticle.

No data on hypothetical modifications in chemical composition of cuticular waxes and cutin monomers in response to the different humidity and temperature conditions applied were reported. It is thus not possible to speculate whether the observed changes in these mechanical characteristics were accompanied by or related partially to compositional alterations in fruit cuticles, or rather that they arose simply from the expectable humidity- and temperature-associated modifications in cuticle structure or physical properties. Nevertheless, these experimental data are also interesting in the light of recent findings that water loss patterns from far less studied fruit species, such as litchi (Litchi chinensis Sonn.) and longan (Dimocarpus longan Lour.) are also biphasic over a wide range of relative humidity values (0%–80%). These two fruit species belong to the Sapindaceae family, some of whose members, among which litchi and longan, do not share the typical pericarp structure of most mature fruits, which display a unique cuticle covering their outer surface. In these fruit species, rather, the whole pericarp develops into a water loss-protective, astomatous structure, which encloses the edible aril. Both the outer and the inner surfaces of this pericarp are covered by cuticles, showing distinctive chemical composition of cuticular waxes and cutin, and representing two in-series resistances against water loss.

This design exemplifies an alternative evolutionary strategy to prevent dehydration of fruits, and illustrates the complexity of water-proofing of aerial plant organs. In both fruit species, exocarp cuticles contained more waxes than endocarp cuticles, and cutin analysis revealed the presence of lignin and suberin monomers in addition to the usual C_{16} and C_{18} cutin components, which may reinforce the transpiration barrier function of cuticles. The mechanistic model developed also revealed that water loss-barrier properties of the exocarp cuticles depended closely on the degree of hydration of the pericarp, which in turn was determined by the external relative humidity, while those of the endocarp were not affected by this environmental factor.

Objectives and Goals of Storage

The main objective of postharvest storage as a component of the value chain is to maintain produce quality as superior as possible for as long as possible. Storage techniques and conditions are important postharvest factors marking quality maintenance of fresh produce. These operations are primarily directed at minimizing the rate of metabolism of harvested produce.

Due to its capital and running expenses, cold storage adds to the cost of production, increasing the price of the produce; the bigger the involvement of the storage system, the higher the added cost. Short-term storage is applied so as to present some marketing flexibility, although it is not worthwhile to store perishables if the price increases, plus storage would lower quality and shelf life. However, the main reasons for the storage of produce are not only associated with marketing, but also with maintaining the quality by considering the following:

- Minimizes decay by slowing down microorganism's progression.
- Lowers transpiration or water losses that otherwise promote unfavorable effects, such as wilting, elongation, rotting, greening, sprouting, and toughening. Such activities affect appearance, quality and texture.

- Slows down the biological activity of fresh produce, such as the case of reducing the production and action of the natural ripening agent ethylene.
- Minimizes the surplus sale in the market, thus guaranteeing good returns to the farmers.
- Assures the accessibility of the produce during the off-season.
- Reduces waste and spoilage of produce.
- Normalizes the price of the produce during the season, as well as during the off-season.

Factors affecting Storage Life

The natural limits to the post-harvest life of all types of fresh produce are severely affected by other biological and environmental conditions:

- Temperature: An increase in temperature causes an increase in the rate of natural breakdown of all produce as food reserves and water content become depleted. The cooling of produce will extend its life by slowing the rate of breakdown.
- Water loss: High temperature and injuries to produce can greatly increase the loss of water from stored produce beyond that unavoidably lost from natural causes. Maximum storage life can be achieved by storing only undamaged produce at the lowest temperature tolerable by the crop.
- Mechanical damage: Damage caused during harvesting and subsequent handling increases the rate of deterioration of produce and renders it liable to attacks by decay organisms. Mechanical damage to root crops will cause heavy losses owing to bacterial decay and must be remedied by curing the roots or tubers before storage.
- Decay in storage: Decay of fresh produce during storage is mostly caused by the infection of mechanical injuries. Furthermore, many fruits and vegetables are attacked by decay organisms which penetrate through natural openings or even through the intact skin. These infections may be established during the growth of the plant in the field but lie dormant until after harvest, often becoming visible only during storage or ripening.

Storage Structures

Ventilated Stores

Naturally ventilated structures can be used for the storage of produce with a long storage potential, such as roots and tubers, pumpkins, onions and hard white cabbage.

Such stores must be designed and built specifically for each intended location. Any type of building can be used provided that it allows the free circulation of air through the structure and its contents.

The following essentials must be observed:

- The building should be located at a site where low night temperatures occur over the required storage period.
- It must be oriented to take maximum use of the prevailing wind for ventilation.
- The material covering the roof and walls should provide insulation from the heat of the sun; grass thatch on a bush-pole frame can be very effective, particularly if it is wetted to provide evaporative cooling.
- Double-skinned walls will provide better insulation, if cost allows.
- White paint applied to surfaces of man-made materials will help to reflect the heat of the sun.
- The structure should be built in the shade of trees if they do not interfere with the prevailing air flow; beware of bush fires and of trees falling during storms.
- Provide ventilation spaces below the floor and between walls and roof to give good air flow.
- If the store is subject to cold night temperatures, fit movable louvres and adjust them to limit the flow of warm air into the store during the day.

These are the basic requirements of a ventilated store. Such stores may be constructed to various levels of sophistication, using, where it is economically acceptable, fan-assisted ventilation controlled by differential thermostats. This type of store is in common use in Europe for the bulk storage of Irish potatoes and onions in locations where external winter conditions make possible the accurate control of the storage temperature.

Simple open-sided, naturally ventilated structures may be used to store seed potatoes at high altitudes in warm climates. They cannot be used for table potatoes, which will turn green, develop a bitter taste, or even become toxic if exposed to light for more than a few hours.

Clamps

These are simple, inexpensive structures used to store root crops, particularly potatoes in Europe and Latin America. The potatoes are placed on a bed of straw I to 3 m wide, but not more than 1.5 m wide in warm climates. A ventilating duct should be placed along the bottom. The piled potatoes are covered with about 20 cm of compacted straw which can subsequently be encased in soil, applied without compaction up to 30 cm deep. The clamp system can be modified for different climatic conditions. In warm climates extra straw casing may be used instead of soil in order to give added ventilation.

Other Simple Storage Methods

Windbreaks are narrow, wire-mesh, basket-like structures about 1 m wide and 2 m high, of any convenient length, on a raised wooden base, and are used for short-term storage of dried onions in the field. The onions are covered on top with a 30 cm layer of straw, which is in turn held down by a polythene sheet fastened to the wire mesh. The windbreak is built at right angles to the prevailing wind to obtain maximum drying and ventilation. Onions can also be woven into plaits on twine and hung in a cool dry place, where they will keep for several months.

Refrigerated and Controlled-atmosphere Storage

For large-scale commercial operations, refrigerated storage may be used in a cold-chain operation to carry regular consignments from production areas to urban markets and retailers. This can be a highly complex operation requiring expert organization and management. Cold storage can also be used for long-term storage of seasonal crops such as potatoes and onions. The storage life of some fruits, such as apples, can be extended by combining refrigeration with a controlled environment consisting of a mixture of oxygen and carbon dioxide. These last are expensive operations with high maintenance and running costs, and demand skilled and experienced management. They have relatively little application to small-scale production in developing countries.

COLD STORAGE

Cold storage facilities are crucial to minimize post-harvest losses; however, losses occur at every step in the post-harvest cycle, and therefore cold storages cannot be considered as independent solutions to prevent post-harvest spoilage but as one component that needs to be integrated in a cold chain network from the point of harvest to the point of purchase by the end consumer.

The different stages in the post-harvest cycle can be summarized as:

- Harvest.
- Precooling.
- Transportation & packaging.
- Cold storage.
- Display at the market.

During these different stages, not only temperature control, but also proper and minimal handling of produce including handling, cleaning, sorting and adequate packing is crucial. Additionally, socioeconomic factors add to the post-harvest loss issue: The lack of marketing systems, transportation facilities and market information are key hindering factors. Also, government regulation, often intended to keep prices low for consumers or to ensure a certain minimum price for farmers, can be counterproductive regarding incentivizing the production of high quality produce. Often, simply the required equipment, such as harvesting tools, containers, cleaning, packing and cooling equipment are not available. Furthermore, the knowledge on post-harvest handling and marketing of farmers and other actors handling the produce, for example during transport, is generally very limited leading to suboptimal treatment of the produce.

Harvest, Handling and Transport

Adequate and proper harvesting and handling of fresh produce are crucial preconditions for long storage and shelf lives of fruit and vegetables. Considering the harvesting and subsequent handling stage of produce, the thumb rule of, the less the better should be followed, in order to minimize chances of mechanical injuries of produce. Once damaged, shelf life is considerably compromised.

Harvesting Principles

For harvest procedures the subsequent points are important to consider:

- Choosing the correct harvest time refers to both the ripeness and maturity of the produce as well as the right time of the day. Optimal harvest times for most produce is either early morning hours or the evening when temperatures are lower.
- Harvesting of delicate, high value produce is best done manually (especially when labor costs are low and fuel costs are high.
- Do not place produce on the ground directly, but use harvesting mats or containers/baskets instead.

Considering the handling of harvested produce, the following points are important:

- Handling:
    - Before being put into storage, produce should be sorted and graded with regards to quality (only high quality produce should enter the storage facility).
    - Produce needs to be cleaned (with clean water in order to avoid the spread of molds and fungi) before being put into storage containers and entering the storage rooms. Dirt bears the potential of introducing pests into the storage facility.

  - Time span between harvest and the placement into storage needs to be kept as short as possible.

- Transport: For the transport of fresh produce, the following points need to be considered:

  - Trucks, carts, etc. Should not be overloaded to avoid damage.
  - Use high quality packaging that will withstand the transport.

Temperature Control - Cold Storage

Fresh food products are characterized by their perishability, and levels of vulnerability to pests and spoilage; however, in general the control of temperature and relative humidity and the prevention of damage can increase shelflife, especially of fresh vegetables and fruits, tremendously. Biological deterioration, caused by respiration rate, ethylene production, mechanical injuries, water stress, physiological disorders and pathological breakdown, leads to decay, loss in nutritive value, and changes in color, texture and flavor. Factors influencing the rate of deterioration are temperature, the level of relative humidity, air velocity, atmospheric composition, i.e. concentration of oxygen, carbon dioxide and ethylene, as well as sanitation.

Molding, pests and other spoilage reasons which lead to quality deterioration, such as loss of water, thrive under warm and humid conditions. Most factors causing product deterioration can be addressed through temperature control, i.e. cooling. A reduction in temperature lowers rates of physiological change (respiration, ethylene production, and enzymatic processes) and slows down the growth of microorganisms (bacteria and fungi). Produce after the point of harvest is still 'alive' meaning that processes of the metabolism, such as respiration, as well as the activity of microorganisms remain ongoing leading to product deterioration. In order to lower respiration rates, delay ripening and to reduce water loss temperature needs to be controlled, either by avoiding the exposure of heat or by creating cooling conditions thereby increasing a product's shelf life. Adequate storage facilities can, to a certain extent, control factors such as temperature, relative humidity and air velocity, increasing product quality, shelf-life and value.

The table below demonstrates the effect of temperate control on shelf life of fresh produce. Temperature reduction, even if kept above the level of optimum storage conditions, can yield tremendous extensions of storage life. Accordingly, any form of cooling, such as moving produce to a shaded and ventilated area or choosing harvest hours in the early morning or evening, is beneficial.

Table: Effect of temperature control on shelf life of fresh agricultural products.

Temperature(°C)	Relative Rate of Deterioration	Relative Shelf Life	Example of Potential Shelf Life in Days
0	1.0	100	45

10	3.0	33	15
20	7.5	13	5
30	15.0	7	2.5
40	22.5	4	1.3

Commodity	Recommended Temperature for Handling and Storage (max post-harvest life)	Post-harvest Life at 35 °C*(ambient temperature)	Post-harvest Life at 25 °C	Post-harvest Life at 15 °C	Increased Marketing Time Available at 15 °C
Cabbage	0 °C (6 Months)	2weeks	4 weeks	8 weeks	4×
Carrots	0 °C (6 Months)	2 weeks	4 weeks	8 weeks	4×
Tomatoes	15 °C (14 days)	3 days	6 days	14 days	5×
Peppers	12 °C (20 days)	3 days	7 days	15 days	5×
Potatoes	5 °C to 7 °C (5 to 10 months)	2 weeks	4 weeks	8 to 10 weeks	4×
Spinach	0 °C (14 days)	1 days	2 days	5 days	5×
Sweet potatoes	15 °C (4 to 6 months)	1 month	2 months	4 to 6 months	4×

After produce has been pre-cooled, it needs to be transported and put into a cold storage facility where, in order to maximize shelf-life and quality, it should be stored at its optimum storage temperature which varies according to different items making the storage of several items in one facility problematic if multiple storage chambers are not available. Storage and shelf life depends on intrinsic characteristic of items stored as well as the storage conditions.

In general, the optimum storage temperature is as low as possible but still above the level of chilling sensitivity of the respective crop. Another important factor affecting shelf-life of fresh produce is the level of relative humidity, impacting water loss, decay, physiological disorders, and the uniformity of ripening. Relative humidity levels should be high and saturated for most types of produce. Finally, ethylene also impacts shelf life: whilst some items are ethylene producers after harvest, other crops are sensitive to ethylene exposure. Subsequently, if produce cannot be stored separately, their requirements regarding temperature, relative humidity and ethylene production/exposure need to be considered.

According to the Postharvest Centre of California, for the sake of short-term storage or transport, three groups of produce can be kept together:

- Most leafy vegetables, cole crops, fruits and berries from temperate climates:0°-2 °C, 90%-98% relative humidity.
- Citrus fruits, most tropical fruits, many fruit-type vegetables: 7°-10 °C, 85%-95% relative humidity.

- Root-type vegetables, winter squashes, most tropical fruits and melons: 13°-18 °C, 85%-95% relative humidity.

Cold Storage Technologies for Small-scale Production

Cooling and cold storage are very energy intensive and expensive undertakings that require reliable electricity supply. This precondition is often not given and capital and running costs for farmers in developing countries are generally too high. Hence, unbroken cold chains and the use of appropriate cold storage facilities are generally not accessible for small-scale farmers. Below, several low technology options are being presented that can operate in electricity scarce regions and are feasible for small-scale farming operations:

Technology	Description	Temperature provided	Fuel
Passive/evaporative cooler	Works in dry, low humidity areas and does not require electricity	10-25 °C	No fuel Architectural measures (shade creation, fountains, etc.)
Absorption refrigerator	Heat-driven technology	below 10 °C	Solar Kerosene
Refrigerators	Electricity-driven and therefore dependent on reliable electricity supply	0 °C	Grid Diesel Renewable sources (hydro, solar, batteries, etc.)

There are multiple factors to consider in the technology and site selection of a cold storage facility:

- Location: The location chosen should be easily accessible and close to market hubs and/or produce collection points.
- Electricity: The access to and the supply of electricity to the cold storage facility should be given. Electricity supply has an impact on the size and equipment feasible of the cold storage.
- Drainage and waste disposal: Good drainage options and waste management equipment needs to be readily available.
- Water: Supply of water is important for matter of sanitation. Also the removal of wastewater generated has to be given.
- Economic viability: The value generated though cold storage needs to justify investment and running costs.

In general, choosing an appropriate cold storage technology depends on the temperatures and level of reliability required, resources available – including skilled labor, capital, electricity and water supply.

Cold Storage Management

Temperature control can have huge benefits for the improvement of shelf-life and product; however, benefits will only realize if the cold storage is managed properly with regards to temperature, relative humidity levels, air circulation, adequate space between storage bins, trays and containers, the mixing of compatible produce, as well as the management of product in and outflow which should follow the 'First In, First Out' principle.

Temperature Management

According to information of the FAO, cold storage facilities should be cooled to the storage temperature required at least three days before produce is being entered.

Chilling Injury

Some produce, especially of tropical origin, is sensitive to chilling, which means that it will incur physiological damages if stored at a certain time period below a certain temperature but above their freezing points. In general, the longer the time period that produce is exposed to temperature below their level of chilling sensitivity and the lower the temperature, the faster damages will occur. It should also be noted that effects can be of a cumulative nature, i.e. the time periods of storage below the level of chilling sensitivity add up even if produce is stored at optimal conditions in between. Several factors, such as the level of maturity and level of ripeness at the point of harvest can affect chilling sensitivity.

Common Visual Symptoms

- Surface lesions (pitting, large sunken areas and discoloration).
- Water soaking (disruption of cell structure and accompanying release of substrate favors the growth of microorganisms).
- Internal discoloration of the pulp.
- Failure to ripen in the expected pattern.
- Accelerated rate of senescence.
- Increased susceptibility to decay.
- Compositional changes.

Freezing Injury

Below 0 °C all type of produce freezes due to dissolved soluble solids which are present in cell saps. Often, damages incurred through freezing only become visible once the produce is returned to temperatures above 0 °C.

Symptoms:

- Waterlogging.
- Glassy areas in the flesh.

Storage Mix

Different produce has different storage demands with regards to factors, such as optimum temperature, level of relative humidity, levels of ethylene production and sensitivity, and chilling sensitivity. Hence, the storage of single commodities is less complex than the storage of multiple commodities; however, the latter can still be the more viable option depending on external circumstances. Besides damages due to mismatches in temperature, also odors can be transmitted and ripening and decay can be affected through the exposure to ethylene, which can result in changes in color, flavor and texture. Further, only high quality produce should be 'allowed' in the cold storage and produce should be sorted accordingly.

Examples of odor transfers which should be avoided:

- Apples/pears with celery, cabbage, carrots, potatoes or onions.
- Celery with onions or carrots.
- Citrus with strongly scented vegetables.
- Pears/apples with potatoes à former acquire unpleasant taste.
- Green pepper will taint pineapples.
- Onions, nuts, citrus, potatoes should be stored separately.

Examples of ethylene producing and ethylene sensitive products:

- Ethylene producing: Example apples, avocado, bananas, pears, peaches, plums, tomatoes.
- Ethylene sensitive produce: Example lettuce, cucumbers, carrots, potatoes, sweet potatoes.

Stacking

Adequate airflow and the even distribution of cold air need to be ensured. It is important to note that air always takes the path where resistance is lowest, and hence partly or unevenly filled storage facilities will have poor cooling rates. The following points should be considered:

- A gap of at least 8 cm between walls and the floor, and the stacks of produce should be kept in order to ensure air flow.
- Well-ventilated storage boxes/containers/crates will improve cooling speed, such as PVC crates or ventilated boxes made of cardboard.
- Adequate space in between storage pallets should be about 4-6 inches.

Operation and Maintenance

Cold storage is a very costly and energy intensive undertaking and therefore the door to the cold storage should be opened as few times as possible and that also in the early morning or evening hours. Also the usage of lighting and fans should be kept at a minimum. Further points should further be considered:

- Logbook: Operators should keep track of incoming and outgoing, as well as produced stored, including anticipated arrival time, quantity, quality, anticipated storage time, storage conditions required, sales price on day stored, actual sales price, energy consumed, etc. of new produce.
- Inventory: An up-to-date inventory should be kept including information about location within cold storage, type, quantity and harvest date of produce, use of pre-cooling method, date produce entered the facility, etc.
- Loading: In general, loading rates should be 4-5% of total capacity. The cold storage should be divided in areas for short-term and longer-term storage space.

Hygiene

In order to prevent pathogen built-up within the facility, strict hygiene measures are pivotal. All hygiene measures have to take necessary safety measures for staff into consideration:

- Clean thoroughly at the end of each season: Wash walls and floor with sanitizers, such as hypochlorite solution.
- Use of ozone generators to prevent molds and fungi.
- Regular ventilation to avoid built-up of ethylene, carbon dioxide and odors.

STORAGE OF FRESH FRUIT AND VEGETABLES

The storage life of fresh fruit and vegetables varies with type, variety and pre-harvest conditions. There is scope to control storage life through postharvest management of the two most important determinants of storage life and quality — respiration and transpiration. Proper control of temperature and relative humidity is the key to maximising storage life and marketable quality.

Harvested horticultural products are living tissues with continuing metabolism after harvest. They are subject to respiration, water loss and cell softening throughout the postharvest system. The storage life of a product varies with species, variety and pre-harvest conditions — particularly quality and maturity. However, there is scope to control storage life through postharvest management of the two most important determinants of storage life and quality — respiration and transpiration. Both need to be limited but not stopped and proper control of temperature and relative humidity is the key to maximising storage life and marketable quality.

Temperature

Each product handled has its own particular temperature requirements. The time until cooling, temperature stability and its duration need to be considered. By lowering produce temperature as soon as possible after harvest — generally within four hours — the following effects are achieved:

- Respiration rate is decreased.
- Water loss is reduced.
- Ethylene production is suppressed.
- Sensitivity to ethylene is reduced.
- Microbial development is slowed.

Most leafy vegetables and 'temperate' fruit — including pome and citrus fruits — are not chill-sensitive and can be stored between 0 °C and 2 °C for long periods without significant loss of visual quality. Tropical and subtropical fruit and some root vegetables are chill-sensitive and may be damaged at low temperatures. They are generally stored at 13 °C or above, although some may be stored safely as low as 5 °C if cooled soon after harvest. Temperature must also be stable as changes may affect respiration and marketable quality.

Relative Humidity

Maintaining high humidity around harvested produce reduces water loss, which would result in decreased returns through poor quality (for example, wilting, shrivelling) and loss of saleable weight. High humidity should be used with low temperature storage because humidity and warmth combined favour the growth of fungi and bacteria. Although the relative humidity of air increases when it is cooled, it is still necessary to check the relative humidity in a cool room is satisfactory.

GRAIN STORAGE

A food grain storage system is an artificial ecological system in which deterioration of the stored product results from interactions among physical (e.g. temperature, moisture, storage structure), chemical (e.g. carbon dioxide, oxygen) and biological (e.g. grain characteristics, microorganisms, insects, mites, rodents, birds) factors. Both pulse grain quality and quantity are affected by intrinsic and extrinsic factors and, among these, temperature and moisture content are the most important influences of shelf-life. Grain viability and reproduction of biological agents in grain are dependent to a great extent on the temperature and moisture levels. Improper storage conditions influence the post-harvest storability of pulses. Various pests and microorganisms attack the pulses and their products after harvest, in storage, during transportation to the market, etc. The application of pesticides results in chemical residues in the food that are extremely hazardous to health. Establishment of efficient and effective post-production storage systems is warranted in order to minimize qualitative and quantitative losses.

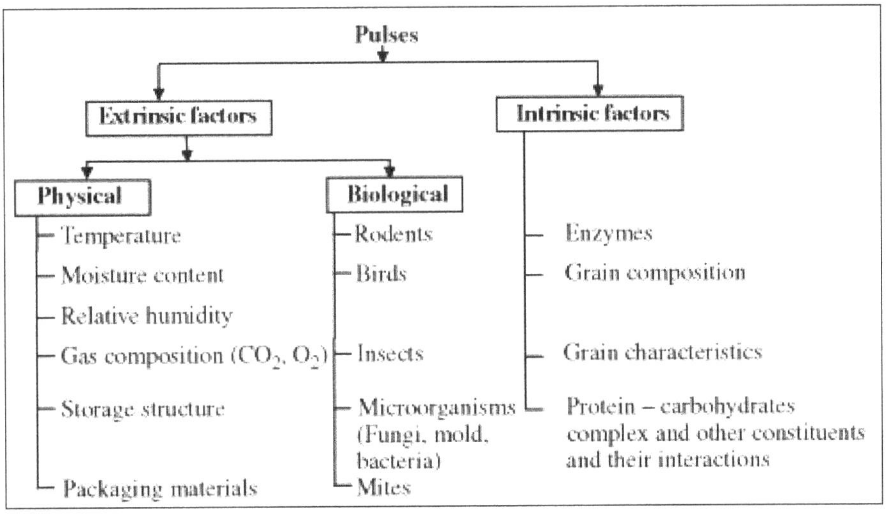

Factors influencing grain quality and quantity.

A storage system is like an ecosystem where quality changes gradually as a result of complex interactions, as shown in figure. Fleurat-Lessard suggested that parameters involved in quality changes can be classified into the following categories:

- Initial condition when grain is delivered to the grain store.
- Storage conditions between harvest and first processing.
- Technical operations and treatments of the grain batch, which is called the storage technical route.
- Biological deterioration factors (also called "deteriorative forces"), mainly represented by storage fungi and invertebrate pests (insects and mites).

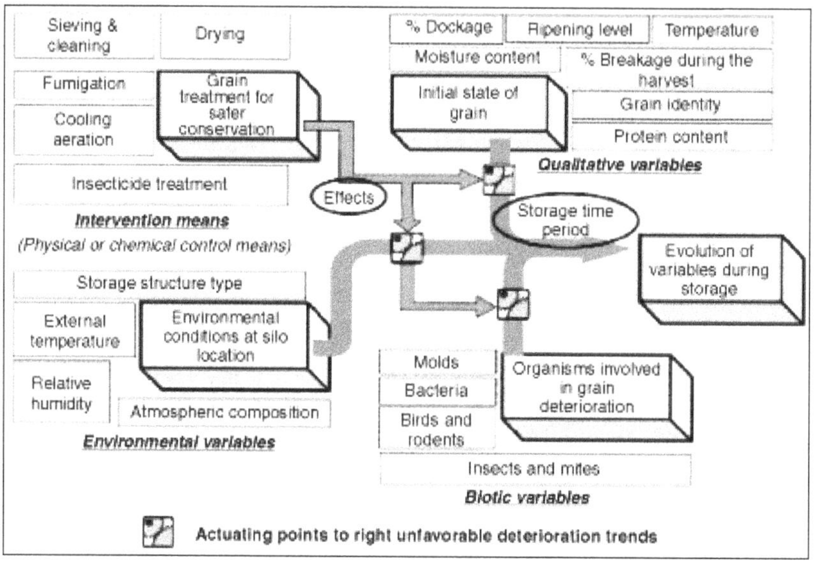

Simplified functional diagram of the complex relationships existing in the stored grain ecosystem.

Storage of pulses is done at several stages of the pulse production chain, mainly on the farm, in warehouses and in other commercial storage facilities at grain processors prior to processing. Both raw whole grains (unprocessed) and processed grains such as dehulled whole, split (dhals) and unhulled splits are stored. The major causes of loss of quality and weight loss in stored grain are the activities of insects, mites, rodents, birds and fungi.

To ensure good storage, grains need to be kept in a dry and clean environment that also protects from pest infestation, foreign material contamination, and unintended grain germination. Compared with high-yielding hybrids, traditional sorghum and millet varieties appear to be more resistant to insect attack. This may be attributable to the fact that early maturing hybrids usually have softer endosperms. Some key factors influencing stored grain quality include grain variety, environment conditions, and storage methods. Both moisture and temperature of storage are also critical physical factors contributing to grain losses. During storage, the relative humidity of the airspace between kernels should be less than 65%. This can be achieved by controlling grain moisture content during drying and temperature of the storage environment by aeration. Grains stored at relative humidity greater than 70% will gradually absorb air moisture and eventually break dormancy. For example, sorghum grains have adsorption and desorption moisture contents of 14.9%–15.8%, 15.9%–16.6%, and 17.0%–17.5%, respectively at 75%, 80%, and 85% relative humidities. Moisture contents greater than the critical 14% will lead to grain invasion by storage fungi. Insects and molds will damage sorghum grain containing 15.5% and 18% moisture content, respectively. Moreover, temperature fluctuations can cause condensation of moisture which can enter the grain. Although grain is initially dried to the recommended moisture level, changes in humidity within the storage vessel during storage and temperature differences within the grain kernels can lead to moisture migration and eventually spoilage.

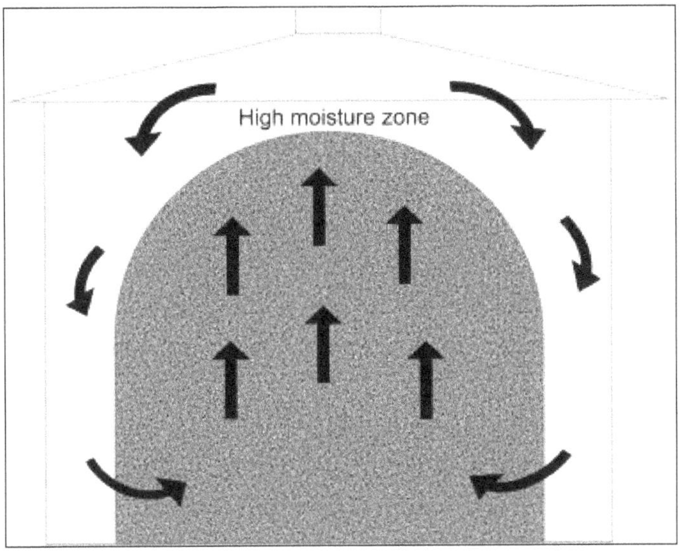

Moisture flow during grain storage.

Sorghum and millets are harvested year round prior to storage as they are grown almost in all seasons of the year. In northern India for example, sorghum is sown in the kharif season (May–July) soon after the first break of monsoon rains, whereas in southern India it is sown in the rabi (September to November) season. Harvesting takes place between October and December and between January and March for the kharif and rabi crops, respectively. Finger millet sowing in India takes place between May and August, and harvesting between September and January. Sowing of pearl millet takes place between May and September, and harvesting between September and February.

Environmental temperatures tend to decrease during storage resulting in cooling of the grain layers close to the walls and roofs of silos and bins. Therefore the central warm and moist air becomes less dense and rises to the cold roof where it condenses and creates a wet favorable zone for fungi and insects on the top of the stored crops. Prevention of such condensation zones within stored grains will discourage growth of fungi and concomitant production of toxins. A well-known poisonous group of toxins produced by fungi are mycotoxins, which include alternariol, patulin, deoxynivalenol, ochratoxin A, zearalenone, and aflatoxins B1, B2 and G1, the maximum levels of which have to comply with the upper limits set by the Codex Committee on Food Additives and Contaminants. Aeration, which is achieved by forcing low volumes of cool air through the stored grain in order to drive out moist air, is therefore essential in maintaining a uniform temperature throughout the mass.

Different storage methods can be used based on grain variety, moisture content, utilization, climate, and storage equipment. Detailed information is provided in the following section, including associated pest control methods.

Silo

Steel grain silos.

Grain elevators are composed of groups of grain silos.

A silo is a structure for storing bulk materials. Silos are used in agriculture to store grain or fermented feed known as silage. Silos are commonly used for bulk storage of grain, coal, cement, carbon black, woodchips, food products and sawdust. Three types of silos are in widespread use today: tower silos, bunker silos, and bag silos.

Types of Silos

Tower Silo

Concrete stave silo under construction.

Storage silos are cylindrical structures, typically 10 to 90 ft (3 to 27 m) in diameter and 30 to 275 ft (10 to 90 m) in height with the slipform and Jumpform concrete silos being the larger diameter and taller silos. They can be made of many materials. Wood staves, concrete staves, cast concrete, and steel panels have all been used, and have varying cost, durability, and airtightness tradeoffs. Silos storing grain, cement and woodchips are typically unloaded with air slides or augers. Silos can be unloaded into rail cars, trucks or conveyors.

Tower silos containing silage are usually unloaded from the top of the pile, originally by hand using a silage fork—which has many more times than the common pitchfork; 12 vs 4—and in modern times using mechanical unloaders. Bottom silo unloaders are utilized at times, but have problems with difficulty of repair. An advantage of tower silos is that the silage tends to pack well due to its own weight, except in the top few feet. However, this may be a disadvantage for items like chopped wood. The tower silo was invented by Franklin Hiram King. In Canada, Australia and the United States, many country towns or the larger farmers in grain-growing areas have groups of wooden or concrete tower silos, known as grain elevators, to collect grain from the surrounding towns and store and protect the grain for transport by train, truck or barge to a processor or to an export port. In bumper crop times, the excess grain is stored in piles without silos or bins, causing considerable losses.

Concrete Stave Silos

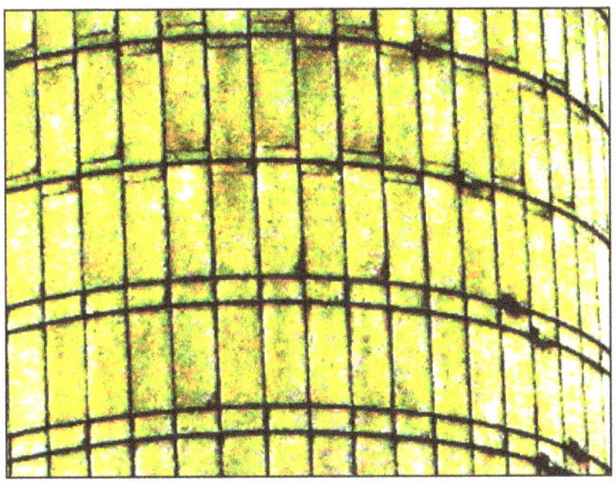

The intermeshed concrete staves, and how the lower hoops are aligned over the stave edges.

Small stave silos can be expanded upward. More hoops are added to strengthen the lower staves.

Concrete stave silos are constructed from small precast concrete blocks with ridged grooves along each edge that lock them together into a high strength shell. Concrete is much stronger in compression than tension, so the silo is reinforced with steel hoops encircling the tower and compressing the staves into a tight ring. The vertical stacks are held together by intermeshing of the ends of the staves by a short distance around the perimeter of each layer, and hoops which are tightened directly across the stave edges.

The static pressure of the material inside the silo pressing outward on the staves increases towards the bottom of the silo, so the hoops can be spaced wide apart near the top but become progressively more closely spaced towards the bottom to prevent seams from opening and the contents leaking out.

Concrete stave silos are built from common components designed for high strength and long life. They have the flexibility to have their height increased according to the needs of the farm and purchasing power of the farmer, or to be completely disassembled and reinstalled somewhere else if no longer needed.

Low-oxygen Tower Silos

Low-oxygen Harvestore tower silos.

Low-oxygen silos are designed to keep the contents in a low-oxygen atmosphere at all times, to keep the fermented contents in a high quality state, and to prevent mold and decay, as may occur in the top layers of a stave silo or bunker. Low-oxygen silos are only opened directly to the atmosphere during the initial forage loading, and even the unloader chute is sealed against air infiltration.

It would be expensive to design such a large structure that is immune to atmospheric pressure changes over time. Instead, the silo structure is open to the atmosphere but outside air is separated from internal air by large impermeable bags sealed to the silo breather openings. In the warmth of the day when the silo is heated by the sun, the gas trapped inside the silo expands and the bags "breathe out" and collapse. At night the silo cools, the air inside contracts and the bags "breathe in" and expand again.

While the iconic blue Harvestore low-oxygen silos were once very common, the speed of its unloader mechanism was not able to match the output rates of modern bunker silos, and this type of silo went into decline. Unloader repair expenses also severely hurt the Harvestore reputation, because the unloader feed mechanism is located in the bottom of the silo under tons of silage. In the event of cutter chain breakage, it can cost up to US$10,000 to perform repairs. The silo may need to be partially or completely emptied by alternate means, to unbury the broken unloader and retrieve broken components lost in the silage at the bottom of the structure.

In 2005 the Harvestore company recognized these issues and worked to develop new unloaders with double the flow rate of previous models to stay competitive with bunkers, and with far greater unloader chain strength. They are now also using load sensing soft-start variable frequency drive motor controllers to reduce the likelihood of mechanism breakage, and to control the feeder sweep arm movement.

Bunker Silos

Bunker silos are trenches, usually with concrete walls, that are filled and packed with tractors and loaders. The filled trench is covered with a plastic tarp to make it airtight. These silos are usually unloaded with a tractor and loader. They are inexpensive and especially well-suited to very large operations.

Bunker silo being filled and compacted.

Bag Silos

Bag silos are heavy plastic tubes, usually around 8 to 12 ft (2.4 to 3.6 m) in diameter, and of variable length as required for the amount of material to be stored. They are packed using a machine made for the purpose, and sealed on both ends. They are unloaded using a tractor and loader or skid-steer loader. The bag is discarded in sections as it is torn off. Bag silos require little capital investment. They can be used as a temporary measure when growth or harvest conditions require more space, though some farms use them every year.

8' diameter by 150 foot silo bag shown just after filling and sealing.

Bins

This silo bin contains 27 variations of stone, sand and gravel.

A bin is typically much shorter than a silo, and is typically used for holding dry matter such as cement or grain. Grain is often dried in a grain dryer before being stored in the bin. Bins may be round or square, but round bins tend to empty more easily due to a lack of corners for the stored material to become wedged and encrusted.

The stored material may be powdered, as seed kernels, or as cob corn. Due to the dry nature of the stored material, it tends to be lighter than silage and can be more easily handled by under-floor grain unloaders. To facilitate drying after harvesting, some grain bins contain a hollow perforated or screened central shaft to permit easier air infiltration into the stored grain.

Cement Storage Silos

Coal silo under construction using aluminum concrete formwork.

There are different types of cement silos such as the low-level mobile silo and the static upright cement silo, which are used to hold and discharge cement and other powder materials such as PFA (Pulverised Fuel Ash). The low-level silos are fully mobile with capacities from 100 to 750 tons. They are simple to transport and are easy to set up on site. These mobile silos generally come equipped with an electronic weighing system with digital display and printer. This allows any quantity of cement or powder discharged from the silo to be controlled and also provides an accurate indication of what remains inside the silo. The static upright silos have capacities from 200 to 800 tons. These are considered a low-maintenance option for the storage of cement or other powders. Cement silos can be used in conjunction with bin-fed batching plants.

Sand and Salt Silos

Sand and salt for winter road maintenance are stored in conical dome-shaped (clear truss roof) silos. These are more common in North America, namely in Canada and the United States. The shaped is based on natural shape formed when piling solids. The dome is made of prefabricated wood panels with singles installed on a circular reinforced concrete base. Open canopy entrance allows for front end loaders to fill and retrieve easily. These are usually found along major highway or key primary roads.

Fabric Silos

Fabric silos are constructed of a fabric bag suspended within a rigid, structural frame. Polyester based fabrics are often used for fabrication of the bag material, with specific attention given to fabric pore size. Upper areas of silo fabric are often manufactured with slightly larger pore size, with the design intent of acting as a vent filter during silo filling. Some designs include metal thread within the fabric, providing a static conductive path from the surface of the fabric to ground. The frame of a fabric silo is typically constructed of steel. Fabric silos are an attractive option because of their relative low cost compared to conventional silos. However, when fabric silos are used to store granular or particulate combustible materials, conventional practices prescribed by established industry consensus standards addressing combustible dust hazards cannot be applied without a considerable engineering analysis of the system.

Forage Silo Usage

Forage Harvesting

Forage silo filling is performed using a forage harvester which may either be self-propelled with an engine and driver's cab, or towed behind a tractor that supplies power through a PTO. The harvester contains a drum-shaped series of cutting knives which shear the fibrous plant material into small pieces no more than an inch long, to facilitate mechanized blowing and transport via augers. The finely chopped plant material is then blown by the harvester into a forage wagon which contains an automatic unloading system.

Tower Filling

Tower forage filling is typically performed with a silo blower which is a very large fan with paddle-shaped blades. Material is fed into a vibrating hopper and is pushed into the blower using a spinning spiral auger. There is commonly a water connection on the blower to add moisture to the plant matter being blown into the silo. The blower may be driven by an electric motor but it is more common to use a spare tractor instead.

A large slow-moving conveyor chain underneath the silage in the forage wagon moves the pile towards the front, where rows of rotating teeth break up the pile and drop it onto a high-speed transverse conveyor that pours the silage out the side of the wagon into the blower hopper.

Bag Filling

Silo bags are filled using a traveling sled driven from the PTO of a tractor left in neutral and which is gradually pushed forward as the bag is filled. The steering of the tractor controls the direction of bag placement as it fills, but bags are normally laid in a straight line. The bag is loaded using the same forage harvesting methods as the tower, but the forage wagon must be moved progressively forward with the bag loader. The loader uses an array of rotating cam-shaped spiraled teeth associated with a large comb-shaped tines to push forage into the bag. The forage is pushed in through a large opening, and as the teeth rotate back out, they pass between the comb tines. The cam-shaped auger teeth essentially wipe the forage off using the steel tines, keeping the forage in the bag.

Before filling begins, the entire bag is placed onto the loader as a bunched-up tube folded back on itself in many layers to form a thick pile of plastic. Because the plastic is minimally elastic, the loader mechanism filling chute is slightly smaller than the final size of the bag, to accommodate this stack of plastic around the mouth of the loader. The plastic slowly unfurls itself around the edges of the loader as the tube is filled.

The contents of the silo bag are under pressure as it is filled, with the pressure controlled by a large brake shoe pressure regulator, holding back two large winch drums on either side of the loader. Cables from the drum extend to the rear of the bag where a large mesh basket holds the rear end of the bag shut. To prevent molding and to assure an airtight seal during fermentation, the ends of the silo bag tube are gathered, folded, and tied shut to prevent oxygen from entering the bag. Removal of the bag loader can be hazardous to bystanders since the pressure must be released and the rear end allowed to collapse onto the ground.

Tower Unloading

A silo unloader specifically refers to a special cylindrical rotating forage pickup device used inside a single tower silo. The main operating component of the silo unloader is suspended in the silo from a steel cable on a pulley that is mounted in the top-center of the roof of the silo. The vertical positioning of the unloader is controlled by an electric winch on the exterior of the silo.

View of silo unloader doors, silage drop tube, and paddle conveyor leading into barn.

For the summer filling of a tower silo, the unloader is winched as high as possible to the top of the silo and put into a parking position. The silo is filled with a silo blower, which is literally a very large fan that blows a large volume of pressurized air up a 10-inch tube on the side of the silo. A small amount of water is introduced into the air stream during filling to help lubricate the filling tube. A small adjustable nozzle at the top, controlled by a handle at the base of the silo directs the silage to fall into the silo on the near, middle, or far side, to facilitate evenly layered loading. Once completely filled, the top of the exposed silage pile is covered with a large heavy sheet of silo plastic which seals out oxygen and permits the entire pile to begin to ferment in the autumn.

Interior view of silo unloader conveyor paddles and drive chain.

In the winter when animals must be kept indoors, the silo plastic is removed, the unloader is lowered down onto the top of the silage pile, and a hinged door is opened on the side of the silo to permit the silage to be blown out. There is an array of these access doors arranged vertically up the side of the silo, with an unloading tube next to the doors that has a series of removable covers down the side of the tube. The unloader tube and access doors are normally covered with a large U-shaped shield mounted on the silo, to protect the farmer from wind, snow, and rain while working on the silo.

The silo unloader mechanism consists of a pair of counter-rotating toothed augers which rip up the surface of the silage and pull it towards the center of the unloader. The toothed augers rotate in a circle around the center hub, evenly chewing the silage off the surface of the pile. In the center, a large blower assembly picks up the silage and blows it out the silo door, where the silage falls by gravity down the unloader tube to the bottom of the silo, typically into an automated conveyor system.

View of doors under shroud. Due to the limited space, the door hinge frame is also the ladder. On the right is the unloader power cable and yellow silage drop tube with removable access doors for insertion of the silage drop spout.

The unloader is typically lowered only a half-inch or so at a time by the operator, and the unloader picks up only a small amount of material until the winch cable has become taut and the unloader is not picking up any more material. The operator then lowers the unloader another half-inch or so and the process repeats. If lowered too far, the unloader can pull up much more material than it can handle, which can overflow and plug up the blower, outlet spout, and the unloader tube, resulting in a time-wasting process of having to climb up the silo to clear the blockages.

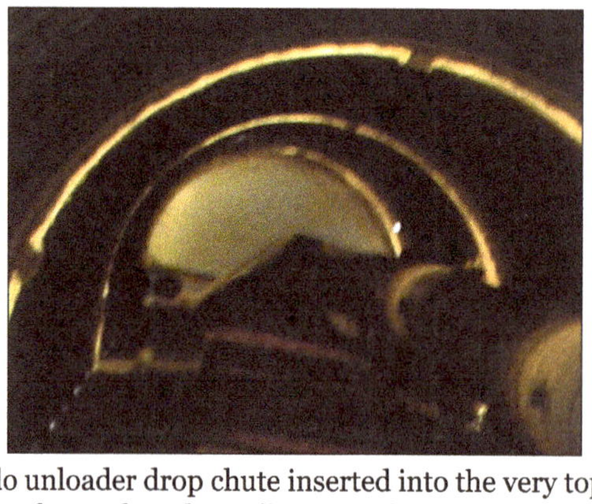

View of the silo unloader drop chute inserted into the very top of the silage drop tube 60ft up. The tube is illuminated by light entering through fiberglass panels every 20ft along the outer steel shroud.

Once silage has entered the conveyor system, it can be handled by either manual or automatic distribution systems. The simplest manual distribution system uses a sliding metal platform under the pickup channel. When slid open, the forage drops through the open hole and down a chute into a wagon, wheelbarrow, or open pile. When closed, the forage continues past the opening and onward to other parts of the conveyor. Computer automation and a conveyor running the length of a feeding stall can permit the silage to be automatically dropped from above by each animal, with the amount dispensed customized for each location.

Safety

Silos are hazardous, and people are killed or injured every year in the process of filling and maintaining them. The machinery used is dangerous and with tower silos workers can fall from the silo's ladder or work platform. Several fires have occurred over the years.

Defunct silo.

Dangers of Loading Process

Filling a silo requires parking two tractors very close to each other, both running at full power and with live PTO shafts, one powering the silo blower and the other powering a forage wagon unloading fresh-cut forage into the blower. The farmer must continually move around in this highly hazardous environment of spinning shafts and high-speed conveyors to check material flows and adjust speeds, and to start and stop all the equipment between loads.

A grain silo.

Preparation for filling a silo requires winching the unloader to the top, and any remaining forage at the base that the unloader could not pick up must be removed from the floor of the silo. This job requires that the farmer work directly underneath a machine weighing several tons suspended fifty feet or more overhead from a small steel cable. Should the unloader fall, the farmer will likely be killed instantly.

Dangers of Unloading Process

Unloading also poses its own special hazards, due to the requirement that the farmer regularly climb the silo to close an upper door and open a lower door, moving the unloader chute from door to door in the process. The fermentation of the silage produces methane gas which over time will outgas and displace the oxygen in the top of the silo. A farmer directly entering a silo without any other precautions can be asphyxiated by the methane, knocked unconscious, and silently suffocate to death before anyone else knows what has happened. It is either necessary to leave the silo blower attached to the silo at all times to use it when necessary to ventilate the silo with fresh air, or to have a dedicated electric fan system to blow fresh air into the silo, before anyone attempts to enter it.

In the event that the unloader mechanism becomes plugged, the farmer must climb the silo and directly stand on the unloader, reaching into the blower spout to dig out the soft silage. After clearing a plug, the forage needs to be forked out into an even layer around the unloader so that the unloader does not immediately dig into the pile and plug itself again. All during this process the farmer is standing on or near a machine that could easily kill them in seconds if it were to accidentally start up. This could happen if someone in the barn were to unknowingly switch on the unloading mechanism while someone is in the silo working on the unloader.

Often, when unloading grain from an auger or other opening at the bottom of the silo, another worker will be atop the grain "walking it down", to ensure an even flow of grain out of the silo. Sometimes unstable pockets in the grain will collapse beneath the worker doing the walking; this is called grain entrapment as the worker can be completely sunk into the grain within seconds. Entrapment can also occur in moving grain, or when workers clear large clumps of grain that have become stuck on the side of the silo. This often results in death by suffocation.

Dry-material/Bin Hazards

There have also been many cases of silos and the associated ducts and buildings exploding. If the air inside becomes laden with finely granulated particles, such as grain dust, a spark can trigger an explosion powerful enough to blow a concrete silo and adjacent buildings apart, usually setting the adjacent grain and building on fire. Sparks are often caused by (metal) rubbing against metal ducts; or due to static electricity produced by dust moving along the ducts when extra dry.

The two main problems which will necessitate silo cleaning in dry-matter silos and bins are bridging and rat-holing. Bridging occurs when the material interlaces over the unloading mechanism at the base of the silo and blocks the flow of stored material by gravity into the unloading system. Rat-holing occurs when the material starts to adhere to the side of the silo. This will reduce the operating capacity of a silo as well as leading to cross-contamination of newer material with older material. There are a number of ways to clean a silo and many of these carry their own risks. However, since the early 1990s acoustic cleaners have become available. These are non-invasive, have minimum risk, and can offer a very cost-effective way to keep a small particle silo clean.

POTATO HOUSE

A potato house is a structure built for the storage of harvested sweet potatoes. Such buildings were common in Sussex County, Delaware, and adjoining areas of Delaware, Maine, and Maryland in the early 20th century, when sweet potato production was at its local peak. Potato houses were typically two-story wood frame structures, of tall and narrow proportions, heated in winter with a coal or wood stove. Potatoes were stored from October through February, requiring a constant temperature of 50 degrees Fahrenheit. The potato houses were constructed to allow the free circulation of air among the stored crop, sometimes employing gaps between the bins and the exterior walls or at the floor level. The tall, narrow proportions of the houses encouraged circulation. Windows were typically provided at the top of the house to allow ventilation if temperatures rose too high. Multiple layers of siding, the use of red rosin paper for sealing against drafts, and sometimes the use of sawdust as insulation were other measures to safeguard the crop.

In general, potato houses were two stories tall, with gable roofs aligned with the long axis. The entrances were through the gable ends, with doors at the first and second floors in the front with an attic window for ventilation in the gables. Rear elevations usually featured a door at the first floor and windows at the second floor and gable. Side elevations were generally plain, with shuttered windows or doors. Openings fit tightly. While two-story elevations were typical, 1-1/2 to three-story houses existed. Most potato houses have been demolished or have been converted to other uses. A number of surviving potato houses has been listed on the National Register of Historic Places.

TECHNIQUES FOR STORING ONIONS

Onions are normally harvested at the end of the bulbing process when "the leaf blades are no longer able to support themselves, the neck softens and the leaves collapse. At this point the bulb has reached maturity. It then enters a period of dormancy during which little change appears to occur." "Onions grow best under dry conditions and at moderate temperatures. Under these conditions onions mature fully and, if they are cured before storing, may last for a long time without loss of quality." Onions grow under very hot or wet conditions face difficulties from fungal and bacterial diseases. In addition to lower yields, the storage losses will probably be higher because bulbs may have been infected before they went into storage. Harvesting after the leaves have completely dried but before the bulbs have begun to resprout, and topping (cutting off the leaves) after most of 59-5 Permanent shed for outdoor drying of onions. The leaf has dried, will decrease storage losses to these plant pathogens. Poor quality "culls" should be removed before the onions are dried and stored. It is best not to promote excessive bulb growth with too much nitrogen fertilizer and irrigation.

Cultivars vary considerably in how well they store. Cultivars suitable for storage should produce a number of outer dry scales or skins which form a vapor barrier around the bulb, thereby minimizing moisture loss and the entry of fungi or bacteria. Locally adapted onion varieties, selected over many years within the tropics, will probably store better than the imported types, especially the 'short-day' varieties from temperate climates.

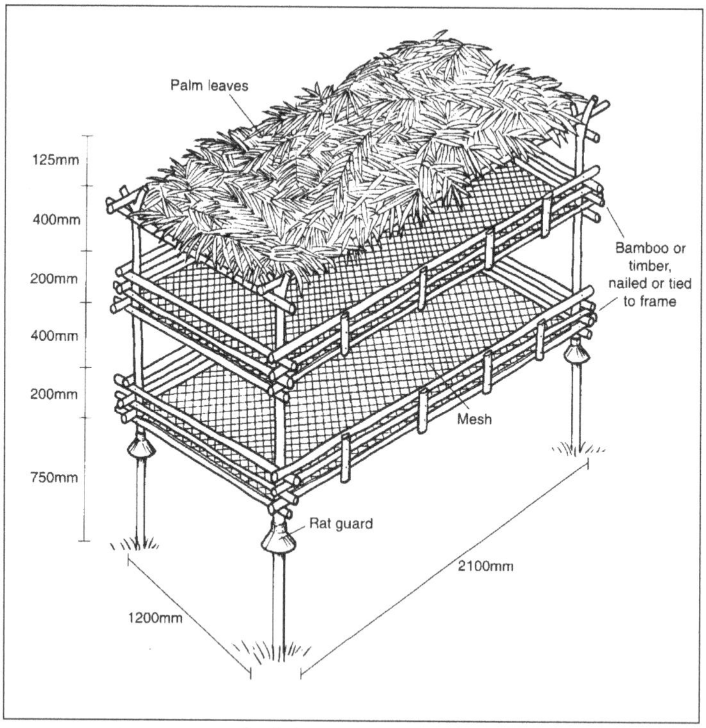

Permanent shed for outdoor drying of onions.

Before storing, bulbs should be properly dried and cured. Use some technique that will both remove surface moisture and allow high temperature formation of strong, intact outer protective skins and neck closure of the onion. High temperatures (e.g. 27 °C or

80 °F), low humidity (< 60% RH) and good ventilation are important in drying. Laying the onions on the soil in windrows is the simplest form of field drying. Alternatively, they may be removed from the field and placed in a heap, or spread in a shallow layer, preferably on a cement slab to avoid contamination of soil-borne pathogens. Problems associated with outdoor drying can include poor drying rates, sun scald, and rain.

A simple outdoor shaded structure can avoid these problems. A good design for natural ventilation is to make shallow trays with screen or woven bottoms. Arrange trays vertically, spaced sufficiently apart to catch the prevailing winds. If clear corrugated plastic sheeting is available, heat required for the curing process can be achieved by using it in the roof structure and thereby capturing passive solar energy.

Maintaining the bulbs at suitable temperature and humidity is vital to the successful storage of onions. The optimum relative humidity range is from 65-75%. There are two favorable temperature regimes. Minimal storage losses occur at 0-5 °C (32-41 °F), but for the farmer or villager not having refrigeration, 25-30 °C (77-86 °F) would be the best choice. Temperatures that are either too high or too low will increase rotting or sprouting. These are the greatest enemies to successful storage of onions. A high technology solution to the sprouting problem is to add chemical "sprout suppressants" prior to storage.

Stored onions should be arranged so that some air reaches each onion, if possible. The storage structure should be designed to maximize natural ventilation. In the tropics there are a variety of ways to do this in home storage. Strings of onions may hang from the ceiling or roof beams or horizontal bamboo rods or racks. In Belize they are hung over the fireplace where wood smoke may have an inhibiting effect on sprouting. They may be stored in bags in the home or in the shade. Ventilation is facilitated by arranging the bags in a particular way. In Pakistan bulbs are sometimes laid on beds of coarse sand. An experiment in the Philippines compared storage of onion layers separated with 50mm of sand, sawdust, rice straw, rice husk or wood shavings. All materials improved storage performance, but sand and sawdust were best.

Many traditional on-farm structures for storing onion are found worldwide. They are constructed from mud, bamboo, straw, wood and other materials with thatching or some fabricated material to protect the interior from rain. A typical good design would be a simple ventilated structure constructed from bamboo. The floor would be at least 15 cm (6 in.) from the ground with framework constructed from poles about 8 cm (3 in.) in diameter placed 0.6 m (2 ft.) apart. The roof is grass thatch and the four sides and base are constructed from split bamboo poles with gaps of 2.5 cm (1 in.) between them to allow good air flow. Onions would be layered within this structure in one of the manners discussed above which will allow air flow to each stored onion. Successful onion storage is the outcome of a process which begins at the crop planning stage and which continues through cultivation, harvesting and post-harvest handling right through to sale of onions to the consumer.

The physical method in which the onions are held (i.e. bulk, tray, box, bin, or bag) and designing the storage area for maximum natural ventilation are important considerations. If power is available, one can make further improvements in the storage conditions by strategically placing fans as a means of forced aeration.

References

- Quinn, Judith (December 23, 1987). "National Register of Historic Places Multiple Property Documentation Form: Sweet Potato Houses f Sussex County, Delaware". National Park Service. Retrieved 25 October 2011
- Postharvest-storage, agricultural-and-biological-sciences: sciencedirect.com, Retrieved 2 March, 2020
- Schneider, Grace (3 May 2018). "The Ultimate Guide To South Australia's Regional Silo Art (& More!)". Glam Adelaide. Retrieved 1 February 2020
- Grain-storage, agricultural-and-biological-sciences: sciencedirect.com, Retrieved 11 August, 2020

5

Postharvest Pathogens and Disease Management

Crops are prone to infectious pathogens after harvest. Such pathogens can also lead to diseases. Common disease causing pathogens include alternaria, sclerotinia, colletotrichum, etc. The topics elaborated in this chapter will help in getting a better perspective about pathogens, the diseases caused by them and their management strategies.

POSTHARVEST DISEASES

The diseases which develop on harvested parts of the plants like seeds, fruits and also on vegetables are the post-harvested diseases. The harvested products may get infected on the way to storage or to market or even before their final consumption. The plant parts may get infected in the field, but expression of symptoms may take place later, at any stage before final consumption.

The plant products may get infected by microorganisms and cause rotting or decaying – partially or totally. The quantity of plant products becomes reduced due to the above infection. The seeds or grains may get damaged by accumulation of toxic substance, the mycotoxin produced by the infected microorganism.

The fleshy fruits, vegetables etc., like tomato, banana, citrus, strawberries, rhizome of zinger, bulb of onion, tuber of potato etc., may get damaged. This results in reduction of quantity, quality or both of the affected parts or products as a whole. The amount or extent of damage depends mainly on the pathogens involved, on the condition of the products and the condition of storage. The pathogens involved are mainly fungi like Pythium, Phytophthora, Rhizopus, Aspergillus, etc. and some bacteria like Pseudomonas, Erwinia, etc.

Classification of Postharvest Diseases

Cristensen and Kaufmann divided the pathogens into two categories:

- Field Pathogen: The field pathogens are those, which cause infection during development of plants or their products before harvest.

- Storage Pathogen: The pathogen which cause infection during storage are the storage pathogen.

Symptoms from infection caused by the 'field pathogens' may be very inconspicuous to be noted at the time of harvest. In fleshy and/or juicy fruits and vegetables, infection by field pathogen continues to develop even after harvest. They may become infected during storage by the same field pathogens or by other pathogens. In seeds and grains, the disease caused by field pathogens ceases to develop further soon after harvest. But, they may be infected further by the other pathogens during storage.

Types of Postharvest Diseases

Stackmann and Harrar divided the pathogenic storage diseases into two groups:

- Diseases of dry, bulk materials, such as seeds and grains.
- Diseases of fleshy storage organs, such as vegetables (tubers, rhizomes, bulbs etc.) and fruits.

Observations of many investigators indicate that the real cause of the spoilage of vegetables and fleshy fruits in transit and also in storage are due to high moisture, high temperature, and injuries caused during marketing. Due to high moisture content and nutrient in harvested vegetables and fruits, they are vulnerable to attack by the pathogenic organisms. Injuries of fruits and vegetables may be caused during harvesting, packing and transposition they help the pathogen to enter the host and cause damage. But the seeds and grains can be stored for long time due to low moisture content (about 12-14%), where most of the pathogens cannot grow favourably.

Diseases of Stored Seeds and Foodgrains

Field fungi, like Alternaria, Fusarium, Cladosporium, Verticillium, Helminthosporium, Colletotrichum etc., attack seeds and grains on growing crops, but are unable to grow in storage due to low relative humidity i.e., below 90%. During storage or transit the seeds and grains are damaged by the different species of Aspergillus and PeniciHum, which can grow well at a relative humidity ranges from 70-90%. The commonly available Aspergillus species are A. repens, A. ruber, A. flavus, A. candidus, etc.

Aspergillus and a number of other storage fungi invade the embryo of the seeds and grains and they discolour the embryo or seeds as a whole, thereby the germination percentage reduces markedly. In some cases, spoilage of stored grains and seeds results in drastic increase of temperature up to 70 °C or more, which encourage the growth of different thermophilic and thermotolerant fungi such as Aspergillus fumigatus, Absidia spp., Mucor pusillus, etc.

In addition to storage fungi, other microorganisms may grow in/or on seeds and accelerate the deterioration process. During breeding period of insects, the moisture content and temperature of seeds increase, thereby rapid growth of the pathogen takes place producing enormous amount of spores. During storage, the fungi produce mycotoxins that cause great damage to both domestic animals and human beings. The important fungi in this respect are Aspergillus and Penicillium, which produce aflatoxin and other toxins.

Most important toxins are:

- Yellow Rice Toxins: Produced in grains of rice, barley etc., by species of Penicillium.
- Tremorgenic Toxins: Produced on prepared food during storage in refrigeration or in other places and also on food produced from infected grains and/or seeds.
- Penicillic Acid: It is a carcinogenic substance produced by the different species of both Aspergillus and Penicillium in molded cereal grains.

Control

The loss or damage by decay and spoilage of seeds and/or grains by storage fungi can be controlled by the following procedures:

- Low Moisture: The moisture content of the rooms for storage should be kept below 70%.
- Low Temperature: Temperature in store house should be maintained below 30 °C, because most of the storage fungi can grow well at temperatures between 30 °C and 55 °C.
- Ventilation: Proper ventilation should be maintained during storage and also during holding period before sending to market.
- Sanitation: Proper sanitation should be maintained to keep storage products clean.
- Use of Insecticide: Insecticides like methyl bromide and some other fumigants are used to treat the harvested seeds, thereby they regulate the storage fungi and reduce economic loss.
- Clean: Clean, uninjured and properly ripened seed should be selected for storage, then only they are able to resist the action of the storage pathogens.

Diseases of Vegetables and Fruits

Different members of Ascomycotina and Deuteromycotina cause the major post-harvest diseases of fruits and vegetables. These are Alternaria, Botrytis, Fusarium, Penicillium, Sclerotinia etc:

- Alternaria: Different species of Alternaria cause rot of many fresh fruits and vegetables, e.g., black rot of orange, tuber rot of potato, rot of sweet potato, purple blotch of onion, Alternaria rot of onion, Alternaria rot of cabbage, etc.

- Botrytis: It causes "grey mold rots" of fruits like pear, apple, citrus etc., and vegetables like onion, tomato etc. Every year it causes great economic loss.
- Fusarium: It causes different diseases, commonly called "pink or yellow molds". Different species of Fusarium cause damage to tubers, bulbs, storage roots etc. and frequently on cucurbits, tomato etc. It also causes brown rot of fruits like lemon, orange etc.
- Penicillium: Species of Penicillium are commonly called "blue or green molds", these cause rots of different fruits like onion, sweet potato etc. They also cause spots on different fruits. Under storage, the spotted fruits bear tufts of spores. Though most of the Penicillium species prefer relatively high temperature for their growth in storage, they still remain active near freezing temperature — at a slow rate. A few species produce ethylene which increases respiration of fruits, thereby it reduces the storage life of the fruits. It also produces patulin — a mycotoxin — which directly contaminates the sauces and fruit juices prepared from infected partly rotten fruits.
- Sclerotinia: It infects different fruits and vegetables. Most common diseases are cottony rot of lemon, watery soft rot of bean pods, cucurbits etc. Storage diseases like bacterial soft rot of vegetables such as onion, carrot, potato etc. are mainly carried out by different species of Erwinia, such as E. carotovora, E. chrysanthemi etc.

Control of Postharvest Diseases

The diseases can be controlled or reduced following the preventive procedures are:

- The fruits and vegetables should be harvested and handled carefully to avoid any injury which may facilitate the pathogen to cause infection.
- The infected region on the vegetables should be cut off to avoid further infection during transportation and storage.
- Storage container, warehouses etc., should be properly cleaned with $CuSO_4$, formaldehyde etc. to avoid contamination.
- The crop should be stored or transported at a temperature low enough to slow down the development of disease.
- Proper ventilation in storage reduces the spread of further development of disease.
- The crops should be free from Insects and other pests, thus creation of new wounds and disease can be avoided.
- Hot water and hot air treatment help to reduce further spread of the disease.

- Chemical control. Post-harvest diseases may be controlled by the application of thiabendazole, dichloran, dosa-ash, etc. These chemicals help to prevent infection and suppress the development of pathogen on the host surface.

Some other chemicals, such as vapours of acetaldehyde, biphenyle/nitrogen chloride forming chemicals etc., are used as supplementary measures to control the post-harvest diseases during storage and transportation.

POSTHARVEST PATHOGENS

Diseases caused by microbial pathogens - fungi, bacteria, and viruses - account for substantial losses of grains, fruits, and vegetables at both pre- and postharvest stages of crop production. The harvested produce might have been infected by pathogens prior to harvest under field conditions or they may get infected during transit and storage. A wide range of fungi and bacteria occurring in the natural environments in which the agricultural and horticultural crops are grown has been screened and selected microorganisms have been employed for the control of the postharvest pathogens.

Infield Penetration

Fungi and bacteria responsible for in-storage decay often originate in the field or orchard. Late blight of potatoes caused by Phytophthora infestans is an example of decay originating in tuber infection in the field. The infection is caused by the zoospores found in soil or that fall onto the tubers from infected foliage during harvest. Following germination, the zoospores penetrate into the tubers through eyes, lenticels, growth cracks, wounds, or via the point of attachment to the plant. Tubers that were infected a few days prior to harvest or during the harvest itself are brought to storage carrying the disease in its early developmental stages, with no visible symptoms of decay.

In a warm and damp climate, the blight fungus can also attack the tomato fruit at its various ripening stages in the field. The attack usually takes place at the edges of the fruit stalk or directly through the skin. In course of epidemics the entire crop may rot in the field. However, when the disease is less severe, fruits with no visible symptoms or with slight blemishes might be picked and the fungus will continue to develop during storage.

The brown rot, which develops in citrus fruits during storage, originates in preliminary infections initiated in the orchards by Phytophthora citrophtliora and other Phytophthora species. During the rainy season, the fungal zoospores descend on the lower fruits of the tree and penetrate them directly. The fungus can develop in the orchard, but zoospore infection occurs a few days prior to picking and the external symptoms are not visible, the fruits will later rot and might comprise a serious problem during storage or transportation.

The principle concern is that the harvest ends fungus development within the host while it is still on the parent plant and causing the decay after harvest or during storage. In many vegetables such as carrot, cucumber, lettuce, celery, cabbage, cauliflower and others, the soft watery rot caused by Sclerotinia sclerotiorum is common. This fungus, common in the soil might attack the vegetable when still in the field or during harvest and continue to develop in its tissues throughout storage or marketing.

Latent Infections

Pathogen penetration takes place in the field and will get to storeroom within the host tissue without eliciting any symptoms of decay known as early or quiescent stages of infection. The fungi that penetrate into the host in fields may cause latent or quiescent infections. These pathogens arrive at fruits or vegetables that are still on parent plant. However, during one of the phases, between their reaching the host and development of progressive disease, the growth is arrested until after the harvest, when physiological and biochemical changes occurring within the host will enable their renewed growth. Verhoeff has described such arrested infections as "latent infections" in which the pathogen growth is temporarily inhibited.

The latent state of the pathogen, whether involving spores that have landed on the host surface have commenced germination, or primary hyphal development within the host tissues, is linked to a dynamic balance among the host, the pathogen and the environment.

The physiological and biochemical changes occurring in the host tissues after and during storage might affect the pathogen and host interrelationship lead to the activation of the latent pathogen. The brown rot, caused by Monilinia fructicola in stone fruits and the decay caused by Nectria galligena in apples originate in the infections of young fruit in orchard.

The Nectria fungus forms a preliminary mycelium within the young tissues, which is the dormant stage of disease, while the interrupted stage in the development of the brown rot caused by Monilinia occur during spore penetration through the stomata or even at the pre-germination stage.

Another example of quiescent infection is anthracnose caused by Colletotrichum gloeosporioides in many tropical and subtropical fruits such as mango,guava, papaya and various citrus fruits. The fungal conidia are found in large quantities on the surface of fruits during their development on the tree. In presence of free water upon the fruit, the spores germinate and form an appressorium at the tip of the germ tube. The appressoria function as resting spores on the fruit, since they resist environmental conditions much better than the conidia and remain vital for long periods while embedded in the natural wax of the fruit or bound to its surface.

Contact Infections

Fruits or vegetables which are free of a pathogen invasion by any means of penetration might be infected through contact with infected produce during storage. Contact infection is a significant factor in spreading white watery rot (Sclerotinia spp.) and bacterial soft rot (Erwinia spp.) in lettuce, carrot, cabbage, celery or squash during storage.

The development of Botrytis infection in stored strawberries, which turns into "mummies" covered with a gray layer of spore-bearing mycelium, causes contact infection and makes fruits susceptible to the disease. Similarly, one strawberry or tomato or a single grape infected by Rhizopus can spread the infection within the container. In fact, contact infection by Botrytis or Rhizopus is typical of many fruits and vegetables and account for major losses caused by these pathogens during long-term storage. In citrus fruits, contact infection of green and blue mould caused by Penicillium digitatum and P. italicum is very common; it often occurs during shipment and under certain conditions disqualifies the entire shipment.

Penetration through Natural Openings

Some pathogenic fungi and bacteria that cannot normally penetrate the directly, without the presence of wound in its surface, can penetrate through natural openings such as stomata and lenticels. The penetration of germ tubes of Colletotrichum gloeosporioides spores into young papaya fruits and the penetration of Monilinia fructicola spore germ tubes into young stone fruits can take place through stomata while the fruit is still in the orchard.

Penetration through lenticels has also been described for Alternaria alternata spores in mango and Gloeosporium perennans in apples. The lenticels in potato tubers are prone to develop bacterial soft rot during storage. Most of lenticels are infested with the cells of Erwinia cartovora during harvest. The bacteria remain inactive in the lenticels until the favourable conditions such as mechanical pressure, existence of free water or low oxygen pressure within the tuber enhancing the tuber sensitivity to decay.

Penetration through Wounds

In contrast to pathogens that attack the fruit and vegetable in the field, most of the storage pathogens are incapable of penetrating directly through cuticle or epidermis of the host, but require a wound or an injury to facilitate their penetration. Therefore, the fungi and bacteria that develop during storage are often called wound pathogens.

Growth cracks present on harvested fruits and vegetables are natural avenues of infections. Careless separation of the fruit or vegetable from the parent plant might result in an injury liable to be attacked by the pathogen. The extent of injury caused by mechanical harvesting is far greater than that caused by a manual operation. The harvesting is accompanied by mechanical injuries enables the weak pathogens to penetrate. It has turned out that the depth of the injury, combined with the humid conditions during storage, determine the fate and extent of the infection.

Each scratch, incision or other mechanical injuries inflicted on the fruit or vegetable during handling, harvesting, transporting, sorting, packing and storing might present adequate penetration points for the storage pathogens. A possible penetration point is the stem end separation area where damage often occurs during fruit picking and in this regard, the separation area is no different from any other injury.

It regularly happens that simultaneously with the injury, a large amount of fungal spores and bacterial cells arrive at the injured area, some of which use the injury site to penetrate and infect the host. Penetration through wounds is a characteristic of Penicillium digitatum and Penicillium italicum pathogen conidia in citrus fruits.

Wound infection is also characteristic of Rhizopus stolonifer that causes watery soft rot in many species of fruits and vegetables; of Alternaria alternata that causes a dark, rather dry decay in a large number of fruits and vegetables stored for a prolonged period; of Geotriclium candidum which causes the sour rot in citrus, melons and tomato fruits; and of various Aspergillus, Cladosporium and Trichothecium species, and other storage fungi of various fruits and vegetables.

Penetration Owing to Result of Senescence

Tissue senescence during continuing storage reduces disease resistance. Commonly at the end of storage period, the sensitivity of melons to blue green mould caused by various species of Penicillium and pink mould caused by Trichothecium roseum is increased. A senescencing onion that has commenced sprouting often harbours base decay, caused by various species of Fusarium. Generally, the rate of decay during storage increases with the duration of storage as tissue senescence progresses. Increasing the tissue sensitivity to diseases during storage also contributes to contact infection of a healthy product by an infected one covered with spore-bearing mycelium.

Penetration after Physical Damage

Injuries caused by low temperature, heat, oxygen shortage or any other environmental stress increases the fruit or vegetable sensitivity to storage fungi. The physiological damage can be externally expressed through tissue browning and splitting, thus forming site vulnerable to invasion of wound pathogens. Extreme environmental conditions enhance sensitivity to an attack without any visible external signs of damage. Development of sun scald lesions in apples leads to Alternaria alternata and Stemphylium hotryosum attack while Alternaria may also be associated with other physiological disorders on apples such as bitter pit or soft scald.

INTEGRATION OF DISEASE MANAGEMENT STRATEGIES

Integration of disease management strategies in fruits and vegetables emerged an answer to shortcomings of reliance on chemical control. It deals with integration of all available disease control methods to manage the diseases effectively and economically to satisfy human needs at the same time guarding the quality of harvested produce and environment. Several studies have highlighted the advantages of Post-harvest application of biological control agents over field or soil application.

The major advantages are the convenience of bringing the antagonist in contact with the commodity as compared to its addition to the soil and the possibility of acting under controlled conditions, created and maintained during storage. Likewise, it is possible to introduce Bacillus subtilis into the wax applied to peaches in the packinghouse to protect them from the brown rot caused by Monilinia fructicola.

The antagonistic yeast Pichia guilliermondii can be introduced into the wax mixture applied to citrus fruit in packinghouse. Compatibility between a microbial antagonist and a synthetic fungicide offers the option of using the antagonist in combination with reduced levels of the fungicide.

Applying the yeast antagonist, Pichia guilliermondii to citrus fruit in combination with substantially reduced concentrations of thiabendazole reduced Penicillium digitatum decay to a level similar to that achieved by the currently recommended concentration of thiabendazole applied alone. Thus, an integrated pest management system adaptation provides effective pest control and maintains very low levels of chemical residues.

The intensive studies on biocontrol of Post-harvest diseases have led to the registration of two biological products for commercial Post-harvest applications on citrus fruits i.e. Aspire, which is Candida oleophila, and Bio-Save 1000 from Pseudomonas syringae.

The biological agent must have low sensitivity to any of the supplemented chemical fungicides. Combining Aspire with each of the chemicals used improved the results; sometimes combinations with a low rate of fungicide were sufficient to achieve effects similar to those obtained by the chemicals at standard rates. Safe compounds or natural products of plant origin have been suggested as alternatives to synthetic and conventional fungicides could also be used in combination with biocontrol agents, complementing their activity. Pathogens treated with such antifungal substances might be weakened and become more vulnerable to the antagonist activity.

The integration of Post-harvest biocontrol into modern production, storage and handling systems must begin before harvest. Several preharvest factors that affect fruit quality may have profound effects on the efficacy of Post-harvest biological control agents.

Preharvest calcium sprays during the growing season of apples and pears can increase fruit firmness, decrease the incidence of certain disorders and enhance resistance to Post-harvest infection. Calcium amendments and Post-harvest application of some antagonistic yeasts can be additive in reducing fruit decay and significantly increase disease control compared with either treatment alone. The advantages in increased firmness, enhanced resistance to Post-harvest decay and enhanced biocontrol efficacy under some circumstances reflect the multiple benefits of integrating Post-harvest biological control with cultural and production practices.

Post-harvest factors may have a major impact on the effectiveness of biological control. Fruit maturity at harvest and the application of antagonists is affecting Post-harvest biological control. Delayed picked and over matured fruits are more susceptible to decay than are fruits picked at optimal storage maturity.

Roberts found that fruit maturity in apples and pears markedly affected biocontrol efficacy of antagonistic yeast Cryptococcus, while excellent control was achieved on freshly harvested fruit and treatments of ripened fruit gave much lower levels of control. Temperature management is a critical factor in the maintenance of fruit quality and pathogen development, may enhance biological control of storage decay.

Janisiewicz demonstrated that as the storage temperature of apples and pears decreased, there was a reduction in the concentration of pyrrolnitrin (a metabolite of Pseudomonas cepacia and other Pseudomonas spp.) needed to protect the fruit from gray mould (Botrytis cinerea) and blue mould caused by Penicillium expansum. An integrated strategy to control Post-harvest decay in pome and stone fruits has been advanced in recent years comprises several pre and Post-harvest components viz., alteration of fruit nutrient status, calcium applications as sprays during the growing season, lower fruit nitrogen content have been associated with reduced disease severity and influence the susceptibility of pome fruits to decay.

Controlled atmosphere storage with reduced O_2 and elevated CO_2 can reduce the severity of Post-harvest fungal decay in apples by inhibiting fruit senescence and maintaining host resistance to infection. Integration of early harvest, low fruit nitrogen, high fruit calcium, yeast or yeast + one-tenth of the dose of thiabendazole along with a controlled atmosphere (2% O_2, 0.6% CO_2) were found to reduce blue mould severity in picked pears. All the components of the integrated approach were found to be compatible with thiabendazole fungicides used for the management of Post- harvest decay and blue mould (P. expansum) in harvested pears.

Several yeast species such as Cryptococcus laurentii, are capable of colonizing wounds of pear fruit under conditions of low temperature, ambient or reduced O_2 and CO_2 therefore, can be integrate into Post-harvest strategies. The two Post-harvest treatments, biological control and modified atmosphere packaging were combined with preharvest iprodione spray; the incidence of the brown rot was reduced from 41.5 per cent in the control to only 0.4 per cent.

A combined strategic approach was elaborated by various workers to control wound infecting pathogens by a series of treatments including disinfection of the fruit surface and environment, eradication or suppression of fungal spore germination at wound sites by a combination of fungicides and reduction of wound susceptibility to infection by the addition of biocontrol antagonists which act as protecting agents.

It is being apparent that a multifaceted integrated approach has been advantageous over the use of solitary measures. Thus, the application of antagonists, induced resistance, physical treatments, chemicals, natural and safe fungicides etc. can be utilized in a unified way to provide greater consistency and efficacy to management of Post-harvest diseases.

POSTHARVEST DISEASES AND THEIR MANAGEMENT IN FRUIT CROPS

Postharvest disease may occur at any time during postharvest handling, from harvest to consumption. Losses caused by postharvest diseases are greater than generally realized because the value of fresh fruits increases several fold while passing from the field to the consumer. Postharvest losses are estimated to range from 10 to 30% per year despite the use of modern storage facilities and techniques. When estimating postharvest disease losses, it is important to consider reductions in fruit quantity and quality, as some diseases may not render produce unsalable yet still reduce product value. Postharvest diseases affect a wide variety of crops particularly in developing countries which lack sophisticated postharvest storage facilities. Infection by fungi and bacteria may occur during the growing season, at harvest time, during handling, storage, transport and marketing, or even after purchase by the consumer. It is also important to take into account costs such as harvesting, packaging and transport when determining the value of produce lost as a result of postharvest wastage. Aside from direct economic considerations, diseased produce poses a potential health risk. A number of fungal genera such as Penicillium, Alternaria and Fusarium are known to produce mycotoxins under certain conditions. Generally speaking, the greatest risk of mycotoxin contamination occurs when diseased produce is used in the production of processed food or animal feed.

Losses due to postharvest disease are affected by a great number of factors including:

- Commodity type,
- Cultivar susceptibility to postharvest disease,
- Postharvest environment (temperature, relative humidity, atmosphere composition, etc.),
- Produce maturity and ripeness stage,
- Treatments used for disease control,
- Produce handling methods,
- Postharvest hygiene.

Virtually all postharvest diseases of fruit are caused by fungi and bacteria. In general, however, viruses are not an important cause of postharvest disease. The so-called 'quiescent' or 'latent' infections are those where the pathogen initiates infection of the host at some point in time (usually before harvest), but then enters a period of inactivity or dormancy until the physiological status of the host tissue changes in such a way that infection can proceed. Examples of postharvest diseases arising from quiescent infections include anthracnose of various tropical fruit caused by Colletotrichum spp. and grey mould of strawberry caused by Botrytis cinerea. The other major groups of postharvest diseases are those which arise from infections initiated during and after harvest.

Postharvest Diseases of Fruit Crops

Fruit crops are attacked by a wide range of microorganisms in the postharvest phase. Actual disease only occurs when the attacking pathogen starts to actively grow in the host. Diseases are loosely classified according to their signs and symptoms. Signs are visible growths of the causal agents, and symptoms the discernible responses produced by the host. Postharvest diseases are caused primarily by microscopic bacteria and fungi, with fungi the most important causal agent in fruit crops. Fungi are further subdivided into classes and are described as lower fungi, characterized by the production of sporangia which give rise to numerous sporangiospores, or higher fungi, described as ascomycetes, deuteromycetes, and basidiomycetes.

Ascomycetes are exemplified by fruiting bodies that release sexual spores when mature. Deuteromycetes, a form of ascomycetes, only release asexual spores. They are more common than the sexual ascomycetes stage in postharvest crops. Deuteromycetes are further subdivided into hyphomycetes and coelomycetes based on spore and structural characteristics. The agonomycetes contain important soil pathogens that form survival structures known as sclerotia that allow them to survive in the absence of the host.

Causes of Postharvest Disease of Fruit Crops

Correct identification of the pathogen causing postharvest disease is central to the selection of an appropriate disease control strategy. Many of the fungi which cause postharvest disease belong to the phylum Ascomycota and the associated fungi Anamorphici. In the case of the Ascomycota, the asexual stage of fungus (anamorph) is usually encountered more frequently in postharvest diseases than the sexual stage of the fungus (teleomorph). Important genera of anamorphic postharvest pathogens include Penicillium, Aspergillus, Geotrichum, Botrytis, Fusarium, Alternaria, Colletotrichum, Dothiorella, Lasiodiplodia and Phomopsis.

In the phylum Oomycota, the genera Phytophthora and Pythium are important postharvest pathogens, causing a number of diseases such as brown rot in citrus (Phytophthora citrophthora and P. parasitica). Rhizopus and Mucor are important genera of postharvest pathogens in the phylum Zygornycota. R. stolonifer is a common wound pathogen of a very wide range of fruit, causing a rapidly spreading watery soft rot. Genera within the phylum Basidiomycota are generally not important causal agents of postharvest disease, although fungi such as Sclerotium rolfsii and Rhizoctonia solani. The major causal agents of bacterial soft rots are various species of Erwinia, Pseudomonas, Bacillus, Lactobacillus and Xanthomonas. Bacterial soft rots are very important postharvest diseases generally of less importance in most fruit.

Host Physiological Status

The development of postharvest disease is intimately associated with the physiological status of the host tissue. To create the right environment for minimizing postharvest losses due to disease it is important to understand the physiological changes that occur after produce is harvested. All plant organs undergo the physiological processes of growth, development and senescence. Growth and development generally only occur while the organ is attached to the plant (with the exception of seed germination and sprouting of storage organs), but senescence will occur regardless of whether the organ is attached or not. When an organ such as a fruit is harvested from a plant, it continues to respire and transpire depleting both food reserves and water. Maturity is a term often used in reference to fruit and is frequently confused with the term ripeness.

Fruit are often classified into two groups on the basis of how they ripen. Climacteric fruit exhibit a pronounced increase in respiration and ethylene production coincidentally with ripening. Climacteric fruit can be harvested in an unripe state and providing they are sufficiently mature, will ripen to an acceptable quality. Non-climacteric fruit do not exhibit a rapid increase in respiration during the ripening process. The eating quality of non-climacteric fruit does not improve after harvest, although they may undergo some changes in colour development and softening. For this reason they should not be harvested until they are ready to eat.

Table: Classification of some common edible fruit is according to respiratory behavior during ripening.

S. No.	Climacteric fruit	Non-climacteric fruit
1	Apple	Blackberry
2	Apricot	Carambola
3	Avocado	Cherry
4	Banana	Grape
5	Blueberry	Grapefruit
6	Custard apple	Lemon
7	Guava	Lime
8	Kiwi fruit	Longan
9	Mango	Litchi

10	Pawpaw (Papaya)	Mandarin
11	Passion-fruit	Orange
12	Peach	Pineapple
13	Pear	Raspberry
14	Plum	Strawberry

Mode of Infection

Infection of fruit by postharvest pathogens can occur before, during or after harvest. Infections which occur before harvest and then remain quiescent until some point during ripening are particularly common amongst tropical fruit crops. Anthracnose, which is the most serious postharvest disease of a wide range of tropical and sub-tropical fruit such as mango, banana, papaya and avocado, is an example of a disease arising from quiescent infections established prior to harvest. Various species of Colletotrichum can cause anthracnose. In avocado for example, early studies reported that ungerminated appressoria were the quiescent phase of C. gloeosporioides. Studies conducted two decades later however showed that appressoria germinated to produce infection hyphae prior to the onset of quiescence. In any case, the fungus ceases growth soon after appressorium formation and remains in a quiescent state until fruit ripening occurs.

In avocados, antifungal dienes are present in the peel of unripe fruit at concentrations inhibitory to Colletotrichum gloeosporioides, the avocado anthracnose pathogen. Grey mould of strawberry caused by Botrytis cinerea is another important postharvest disease sometimes arising from quiescent infections established before harvest. Conidia of B. cinerea on the surface of necrotic flower parts germinate in the presence of moisture. The fungus colonises the necrotic tissue and then remains quiescent in the base of the floral receptacle. Many postharvest diseases develop from the stem end of fruit. The mode of infection involved in this group of diseases can however vary considerably. In the example of B. cinerea, lesions occurring at the stem end of fruit arise from quiescent floral infections. Stem end rots of citrus caused by Lasiodiplodia theobromae and Phomopsis citri result from quiescent infections in the stem button of fruit. In other stem end rot diseases, infection occurs during and after harvest through the wound created by severing the fruit from the plant (e.g. banana crown rot). Endophytic infection, where by the fungus symptomlessly and systemically colonises the stem, inflorescence and fruit pedicel tissue, is important in a number of stem end rot diseases of tropical fruit. Mango stem end rot caused by Dothiorella dominicana is one example of a postharvest disease arising from endophytic colonisation of fruit pedicel tissue. In this case, the fungus colonises the pedicel and stem end tissue of unripe fruit, where it remains quiescent until fruit ripening commences.

Postharvest diseases which can arise from late season infections include brown rot of peach (Monilinia fructicola) and grey mould of grape (Botrytis cinerea). Mechanical injuries such as cuts, abrasions, pressure damage and impact damage commonly occur during harvesting and handling. Some chemical treatments used after harvest, such as fumigants used in insect disinfestations and disinfectants such as chlorine, may also injure produce if applied incorrectly. Tropical fruit in particular are very sensitive to low temperatures, many developing symptoms of chilling injury below 13 °C (depending on storage duration). For example, the incidence of alternaria rot in pawpaw and apple is increased by exposure to excessive cold. For example, hot water dipping of mangoes for excessive times or temperatures can result in increased levels of stem end rot.

Prcharvcst Factors that Influence Postharvest Diseases

- Weather: Weather affects many factors related to plant diseases, from the amount of inoculum that overwinters successfully to the amount of pesticide residue that remains on the crop at harvest. Abundant inoculum and favorable conditions for infection during the season often result in heavy infection by the time the produce is harvested. For example, conidia of the fungus that causes bull's-eye rot are rain dispersed from cankers and infected bark to fruit especially if rainfall is prolonged near harvest time, causing rotten fruit in cold storage several months later.
- Physiological Condition: Condition of produce at harvest determines how long the crop can be safely stored. For example, apples are picked slightly immature to ensure that they can be stored safely for several months. The onset of ripening and senescence in various fruits renders them more susceptible to infection by pathogens. On the other hand, fruits can be made less prone to decay by management of crop nutrition.
- Fungicide Sprays: Certain pre-harvest sprays are known to reduce decay in storage. Several studies have been done on the effectiveness of pre-harvest ziram fungicide application on pome fruit and show an average reduction in decay of about 25 to 50% with a single spray. Iprodione has been used for several years as a pre-harvest spray 1 day before harvest to prevent infection of stone fruit by Monilinia spp. In combination with wax and/or oil its decay control spectrum is increased and it will also control postharvest fungi such as Rhizopus, and Alternaria. The new class of strobilurin fungicides promises to provide postharvest control of several diseases in fruits. They are especially effective against fruit scab on apples and should reduce the presence of pin point scab in storage.

Postharvest Factors that Influence Decay

Packing Sanitation

It is important to maintain sanitary conditions in all areas where produce is packed. Organic matter (culls, extraneous plant parts, soil) can act as substrates for decay-causing pathogens. For example, in apple and pear packinghouses, the flumes and dump tank accumulate spores and may act as sources of contamination if steps are not taken to destroy or remove them. Chlorine readily kills microorganisms suspended in dump tanks and flumes if the amount of available chlorine is adequate. A level of 50 to 100 ppm of active chlorine provides excellent fungicidal activity. Chlorine measured as hypochlorous acid can be obtained by adding chlorine gas, sodium hypochlorite, or dry calcium hypochlorite. Although chlorine effectively kills spores in water it does not protect wounded tissue against subsequent infection from spores lodged in wounds. Organic matter in the water inactivates chlorine, and levels of chlorine must be constantly monitored. Recently, in precisely controlled tests in water or as foam, chlorine dioxide was found to be effective against common postharvest decay fungi on fruit packinghouse surfaces.

Postharvest Treatments

Products used for postharvest decay control should only be used after the following critical points are considered type of pathogen involved in the decay:

- Location of the pathogen in the produce.
- Best time for application of the treatment.
- Maturity of the host.
- Environment during storage, transportation and marketing of produce. Specific materials are selected based on these conditions and fall into either chemical or biological categories.

Fungicide Treatments

Several fungicides are presently used as postharvest treatments for control of a wide spectrum of decay-causing microorganisms. However, when compared to preharvest pest control products the number is very small. For example, intensive and continuous use of fungicides for control of blue and green mold on citrus has led to resistance by the causal pathogens of these diseases. Chemical treatments that are presently used are thiabendazole, dichloran, and imazalil. However, resistance to thiabendazole and imazalil is widespread and their use as effective materials is declining. These products include sodium benzoate, the parabens, sorbic acid, propionic acid, SO_2, acetic acid, nitrites and nitrates, and antibiotics such as nisin.

Biological Control

Postharvest biological control is a relatively new approach and offers several advantages over conventional biological control:

- Exact environmental conditions can be established and maintained.

- The bio-control agent can be targeted much more efficiently.
- Expensive control procedures are cost-effective on harvested food.

Several biological control agents have been developed in recent years, and a few have actually been registered for use on fruit crops. The first biological control agent developed for postharvest use was a strain of Bacillus subtitles. It controlled peach brown rot, but when a commercial formulation of the bacterium was made, adequate disease control was not obtained.

Irradiation

Although ultraviolet light has a lethal effect on bacteria and fungi that are exposed to the direct rays, there is no evidence that it reduces decay of packaged fruits. More recently, low doses of ultraviolet light irradiation (254 nm UV-C) reduced postharvest brown rot of peaches. In this case, the low dose ultraviolet light treatments had two effects on brown rot development; reduction in the inoculum of the pathogen and induced resistance in the host. Gamma radiation has been studied for controlling decay, disinfestation, and extending the storage and shelf-life of fresh fruits. A dose of 250 Gy has an adverse effect on grapefruits increasing skin pitting, scald, and decay. Low doses of 150 for fruit flies and 250 gray (Gy) for codling moth are acceptable quarantine procedures. Gamma irradiation may be used more in the future once methyl bromide is no longer available to control insect infestation in stored products. All uses of methyl bromide are being phased out to avoid any further damage to the protective layer of ozone surrounding the earth.

Temperature and Relative Humidity

Proper management of temperature is so critical to postharvest disease control that all other treatments can be considered as supplements to refrigeration. Fruit rot fungi generally grow optimally at 20 to 25 °C (68 to 770F) and can be conveniently divided into those with a growth minimum of 5 to 10 °C (41 to 50 °F) or -6 to 0 °C (21.2 to 32 °F). Fungi with a minimum growth temperature below-2 °C (28.4 °F) cannot be completely stopped by refrigeration without freezing fruit. High temperature may be used to control postharvest decay on crops that are injured by low temperatures such as mango and papaya. Heating of pears at temperatures from 21 to 38 °C (69.8 to 100.4 °F) for 1 to 7 days reduced postharvest decay. Decay in "Golden Delicious" apples was reduced by exposure to 38 °C (100.4 °F) for 4 days and virtually eliminated when treated after inoculation.

Modified or Controlled Atmospheres

Alterations in O_2 and CO_2 concentrations are sometimes provided around fruit and vegetables. Because the pathogen respires as does produce, lowering the O_2 or raising the CO_2 above 5% can suppress pathogenic growth in the host. In crops such as stone fruits, a direct suppression occurs when fungal respiration and growth are reduced by the high CO_2 of the modified atmosphere. Low O_2 does not appreciably suppress fungal growth until the concentration is below 2%. Important growth reductions result if the O_2 is lowered to 1% or lower although there is a danger that the crop will start respiring an aerobically and develop off-flavor.

Integrated Control of Postharvest Diseases

Effective and consistent control of storage diseases is dependent upon integration of the following practices:

- Select disease resistant cultivars where possible.
- Maintain correct crop nutrition by use of leaf and soil analysis.
- Irrigate based on crop requirements and avoid overhead irrigation.
- Apply pre-harvest treatments to control insects and diseases.
- Harvest the crop at the correct maturity for storage.
- Apply postharvest treatments to disinfest and control diseases and disorders on produce.
- Maintain good sanitation in packing areas and keep dump water free of contamination.
- Store produce under conditions least conducive to growth of pathogens.

Integration of cultural methods and biological treatments with yeast biocontrols has been studied on pears. These results demonstrated that unrelated cultural and biological methods that influenced pear decay susceptibility can be combined into an integrated program to substantially reduce decay. As a general rule, alternatives to chemical control are often less effective than many fungicides. Therefore, it will generally be necessary to combine several alternative methods to develop an integrated strategy to successfully reduce postharvest decay.

Postharvest Losses and their Management

Postharvest disease, disorders or losses in quality have economic impacts vastly greater than the actual losses caused by frequency and intensity of their occurrence. For example there are direct financial losses incurred by the grower from batches of fruit expressing the disease and disorder. Direct losses can also cause financial losses for postharvest operators and marketers. Another point to remember is that the loss of value of a downgraded product is likely to be substantially greater for highly differentiated branded products which sell at a premium in the market. All the hard work that has gone into promoting and raising the profile of a branded product can be quickly eroded if there are postharvest quality problems with some lines of that product.

- Management: Horticultural produce is alive and has to stay alive long after harvest. Like other living material it uses up oxygen and gives out carbon dioxide. It also means that it has to receive intensive care. For a plant, harvesting is a kind of amputation. In the field it is connected to roots that give it water and leaves which provide it with the food energy it needs to live. Once harvested and separated from its sources of water and nourishment it must inevitably die. Horticultural managers must possess many skills to succeed in this. They need a keen appreciation of horticultural diversity. For example, apples and bananas each have their own requirements.
- Harvest handling: The care taken during harvesting is repaid later, because fewer bruises and other injuries mean less disease and enhanced value. Good managers train their pickers so that they select the product at the correct stage of maturity with adequate care. It is worthwhile reducing the amount of hard physical work required in picking fruit as far as possible.
- Pre-cooling: The harvested produce has to be transported to the packing shed without delay. In the field the heat of the sun and the respiration of the produce combine to heat up the produce, especially in the centre of field bins. This accumulation of "field heat" reduces the postharvest life of the product and has to be removed quickly. Strawberries for example, respire nearly eight times faster at a field temperature of 25 °C as they do in a storage temperature of 0 °C. Precooling requires a greater refrigeration capacity than doe's cool storage and is often best done as a separate step. Hydrocooling with cold water drenches, forced air cooling through stacks that ensure proper air distribution and packing with ice are the systems most commonly used, with the choice depending on the individual requirements of the commodity.
- Natural fungicides: Many compounds produced naturally by microorganisms and plants have fungicidal properties. Chitosan, for example, is not only an elicitor of host defence responses but also has direct fungicidal action against a range of postharvest pathogens. Antibiotics produced by various species of Trichoderrna have potent antifungal activity against Botrytis cinerea, Sclerotinia sclerotiorum, Corticium rolfsii and other important plant pathogens. There are many other natural compounds which have been isolated and shown to possess considerable antifungal activity.
- Postharvest fungicides can be applied as dips, sprays, fumigants, treated wraps and box liners or in waxes and coatings. Dips and sprays are very commonly used and depending on the compound, can take the form of aqueous solutions, suspensions or emulsions. Fungicides commonly applied as dips or sprays include the benzimidazoles (e.g. benomyl and thiabendazole) and the triazoles

(e.g. prochloraz and imazalil). The benzimidazole groups of fungicides are very useful for the control of many important postharvest pathogens such as Penicillium and Colletotrichum, Fumigants, such as sulphur dioxide for the control of grey mould (Botrytis cinerea) of grape and various postharvest diseases of litchi, are sometimes used for disease control. Other fumigants used in certain situations include carbon dioxide, ozone and ammonia. Fruit wraps or box liners impregnated with the fungicide biphenyl are used in some countries for the control of Penicillium in citrus. Fungicides to control postharvest diseases of citrus and some other fruit are often applied to the fruit in wax on the packing Iine.

- Refrigeration: Refrigeration is the most important tool for extending the life of fruit. In a typical cool store. To maintain a storage temperature of 0 °C the temperature of the coils will have to be appreciably below 0 °C. Moisture is therefore removed from the air and this accumulates as ice on the coils. The lower the average temperature of the cooling coils, the more moisture will be removed. The drier and cooler air then circulates around the room where it warms and picks up moisture. The more moisture that freezes on the refrigerator coils, the greater the frequency of defrosts cycles and these make good temperature management control more difficult to attain. An even distribution of air produces a room with a consistent temperature, but if the flow of cooled air is "short circuited" back to the coiling coils, the areas starved of circulation will become warmer.
- Quality control: Most consumers have been disappointed with the quality of fresh produce they have purchased at one time or another. Fruit are to be stored or transported over long distances may have to be picked in an immature state so that the fruit are firm and store or travel well. In recent years, much work has been done to improve the quality of fruit. New varieties have been introduced which gives consumers a wider choice and some, such as new varieties of nectarine have improved flavor. However, despite the best efforts to handle fresh produce in the optimum way there will always be compromises to be made which affect the final quality of the product. Much of the fresh produce we eat must travel many kilometers to reach the central or wholesale market or distribution centre and from there it travels to local or distant retail outlets. Good handling will ensure that the final consumers are satisfied and so will return again to buy that product.

CHLORINATION AND POSTHARVEST DISEASE CONTROL

In a hungry and increasingly competitive world, reducing postharvest food losses is a major agricultural goal. For highly perishable commodities, such as tomatoes, squash, and peaches, as much as 30 percent of the harvested crop may be lost to postharvest diseases before it reaches the consumer. Investments made to save food after harvests are usually less costly for the grower and the consumer and less harmful to the environment than efforts to increase production. Even a partial reduction in postharvest losses can significantly reduce the overall cost of production and lessen our dependence on marginal land and other scarce resources.

Many factors contribute to postharvest losses in fresh fruits and vegetables. These include environmental conditions such as heat or drought, mechanical damage during harvesting and handling, improper postharvest sanitation, and poor cooling and environmental control. Efforts to control these factors are often very successful in reducing the incidence of disease. For example, reducing mechanical damage during grading and packing greatly decreases the likelihood of postharvest disease because many disease-causing organisms (pathogens) must enter through wounds.

Chemicals have been widely used to reduce the incidence of postharvest disease. Although effective, many of these materials have been removed from the market in recent years because of economic, environmental, or health concerns.

Increased interest in the proper postharvest handling of fresh fruits and vegetables in North Carolina has prompted the widespread use of flumes, water dump tanks, spray washers, and hydrocoolers. To conserve water and energy, most postharvest processes that wet the produce recirculate the water after it has passed over the produce. This recirculated water picks up dirt, trash, and disease-causing organisms. If steps are not taken to prevent their spread, these organisms can infect all the produce that is subsequently processed. In the past, various fungicides and bactericides have been used (alone or in combination with chlorination) to prevent the transmission of diseases. These materials have often been favored over chlorination because they provide some residual protection after treatment.

At present, chlorination is one of the few chemical options available to help manage postharvest diseases. When used in connection with other proper postharvest handling practices, chlorination is effective and relatively inexpensive. It poses little threat to health or the environment. This publication has been prepared to acquaint growers, packers, and shippers with the proper use of chlorination.

Chemistry of Clorination

Chlorine is a very irritating, heavy, greenish yellow gas with a strong, pungent odor. Free chlorine is very reactive, combining with any chemical that will react with oxygen, and is never found uncombined in nature. Chlorine in the gaseous form is a very potent disinfectant, although it is seldom used in that form. It is much safer and easier to use when dissolved in water. Disinfection of produce using chlorine or some other chemical is nearly always done during hydrocooling or during the process of washing the produce to remove soil. Chlorine for disinfection may be obtained from one of three sources: pressurized chlorine gas, calcium hypochlorite (a soluble solid), or a solution of sodium hypochlorite.

Chlorine Gas

Chlorine gas is produced by the electrolysis of salt solutions (principally NaCl) and is furnished commercially in pressurized metal cylinders. Chlorination is accomplished by bubbling a metered amount of the gas into the supply water. Because of dangers involved with the use of chlorine gas and the expense of the metering equipment, the use of the gaseous form for chlorination is usually limited to large applications. Most municipal water supplies are disinfected with chlorine gas. The limited amount of chlorination required by most postharvest fruit and vegetable operations makes the use of chlorine gas impractical.

Calcium Hypochlorite

The most common source of chlorine used in postharvest chlorination is calcium hypochlorite. It is available commercially in the form of either a granulated powder or large tablets. Most commercial formulations are 65 percent calcium hypochlorite, with the balance consisting of stabilizers and inert materials. Calcium hypochlorite is relatively stable as long as it is kept dry, and it may be stored for extended periods. The property that makes it stable also makes it difficult to dissolve completely in water. Adding granulated calcium hypochlorite directly to the water often results in undissolved particles that adhere to the produce, causing undesirable bleaching and chlorine burns. This problem is particularly common in hydrocoolers because calcium hypochlorite is very slow to dissolve in cold water. Therefore, always dissolve granulated calcium hypochlorite in a small quantity of tepid water before adding it to the wash tank or hydrocooler. Calcium hypochlorite may be obtained in tablets that are added directly to the hydrocooler or wash tank to eliminate the problem of chlorine burns. Properly used, the tablets will dissolve slowly to yield a continuous supply of chlorine to the water. However, the tablets must be positioned carefully to ensure proper mixing of the chemical with the water.

Sodium Hypochlorite

The active ingredient of most liquid household bleaches, sodium hypochlorite is commonly used when the scale of postharvest chlorination is limited. Sodium hypochlorite is not generally available in solid form because it is difficult to store. It absorbs moisture readily from the atmosphere, causing it to release chlorine gas.

Household bleach is usually marketed as a solution of water and 5.25 percent sodium hypochlorite. Larger containers of 12.75 percent or 15 percent sodium hypochlorite solutions are also available through some laundry and swimming pool chemical suppliers. For the same amount of chlorination, a sodium hypochlorite solution is generally more expensive than granular calcium hypochlorite because of the additional shipping and handling costs associated with the water it contains.

Chlorination chemicals can be added to the water manually, or concentrated solutions of sodium or calcium hypochlorite can be injected into the wash tank or hydrocooler at a continuous and measured rate. Commercial chlorine injector systems are particularly useful in operations where a continuous supply of clean, chlorinated water is required. Injector systems consist of a feed tank and an electrically operated pump with a variable output. Chlorine injectors should always be isolated from water supply lines with an approved check valve to prevent backflow into the fresh water system.

Chlorination Effectiveness

Chlorination is a dynamic chemical process. Its effectiveness is influenced by a number of factors. Proper chlorination requires frequent monitoring of the solution and a thorough understanding of the factors involved. These factors include the pH of the solution, chlorine concentration, water temperature, amount of organic matter present, exposure time, and the growth stage of the pathogens present.

Solution pH

The pH of a solution is a measure of its acidity or alkalinity. A solution that is neutral (neither acid nor alkaline) has a pH of 7.0. Solutions with pH numbers less than 7.0 are acid; the lower the number, the greater the acidity. On the other hand, the greater the number above 7.0, the more alkaline the solution. A change of one pH unit indicates a ten-fold change in acidity or alkalinity.

The pH of the solution has a significant effect on the level of chlorination activity. When chlorine gas or one of the hypochlorite salts is added to water, each will generate chlorine gas (Cl_2), hypochlorous acid (HOCl), or hypochlorite ions (OCl) in various proportions, depending on the pH of the solution. The form desired for chlorination is hypochlorous acid (HOCl). Hypochlorite ions are relatively inactive, and chlorine gas quickly bubbles out of the solution, causing worker discomfort and serving no useful purpose.

At a pH slightly above neutral, half of the chlorine will be in the form of hypochlorous acid and the other half in the form of hypochlorite ions. Very little will be in the gaseous form. Solutions that are more acid have a higher percentage of hypochlorous acid but are very unstable, allowing more of the chlorine to escape from the solution as a gas. To maximize the proportion of hypochlorous acid and hence the effectiveness of the solution, the pH should be kept in the practical range between 6.5 and 7.5.

Because well water in North Carolina varies from moderately acid to moderately alkaline, the pH should be checked with a pH meter or test papers before and after the chemicals are added and frequently during operation. Furthermore, even if the water initially has a near-neutral pH, the addition of hypochlorites will change the pH.

Different sources of chlorine have different effects on pH:

- Chlorine gas decreases pH.

- Sodium hypochlorite increases pH.
- Calcium hypochlorite increases pH slightly.

It may be necessary to add a common acid (like vinegar) to lower the pH. Small amounts of sodium hydroxide (lye) may be used to raise the pH. Inexpensive test papers for checking both the chlorine level and pH may be obtained from most swimming pool and chemical supply houses.

Chlorine Concentration

The concentration of a small amount of chemical in a solution is measured in units of parts per million (ppm). In the case of chlorination, this unit of measure indicates the number of parts of available chlorine, by weight, that there are in a million parts of solution. The quantity of calcium or sodium hypochlorite that must be added to a certain quantity of water to obtain a given concentration depends on:

- The available chlorine content of the compound.
- The concentration of the compound.
- The volume of water to be treated.

Table below shows the minimum chlorine concentrations needed to kill all pathogens within one minute at two different temperatures, assuming a neutral pH. Table below gives the amount of 5.25 percent solution hypochlorite solution that must be added to 100 gallons of water to obtain various chlorine concentrations from 25 to 150 ppm. Table gives the same information fo 65 percent calcium hypochlorite granules.

Table: Minimum chlorine concentration necessary to kill all pathogens within 1 minute at two temperatures at neutral pH.

	Chlorine Concentration (ppm)	
	77 °F	104 °F
Fungi	30-40	10
Bacteria	20	10

Table: Amounts of 5.25 percent sodium hypochlorite (NaOCl) solution required to obtain a specified concentration of chlorine in 100 gallons of water at neutral pH.

Pints of solution per 100 gallons of water	Approximate chlorine concentration (ppm)
0.4	25
0.8	50
1.2	75

1.6	100
2.0	125
2.4	150

Table: Amount of 65 percent calcium hypochlorite ($Ca(OCl)_2$) granules required to obtain a specified concentration of chlorine in 100 gallons of water at neutral pH.

Ounces of granules per 100 gallons of water	Approximate chlorine concentration (ppm)
0.5	25
1.0	50
1.5	75
2.0	100
2.5	125
3.0	150

Temperature

The activity of chlorine increases with the temperature of the solution. Unnecessarily warm solutions should be avoided, however, because the chlorine escapes into the air more rapidly as the temperature increases. On the other hand, in hydrocooling the combined effects of low temperature and high pH values reduce chlorination efficiency.

Organic Matter

Chlorine has a particular affinity for soil particles and organic matter. Chlorinating dirty produce therefore depletes the chlorine supply much faster than relatively clean produce. The amount of chlorine constantly decreases with chlorination reactions. The more organic matter (such as fruit, leaves, or soil) in the tank, the more chlorine will be lost. As a result, the chlorine level should be checked and adjusted hourly, especially when large loads of produce are being processed. Extremely dirty produce (such as sweetpotatoes) is commonly washed with clean water before it is placed into the chlorination tank.

Exposure Time

The effectiveness of chlorination depends greatly on the length of time the produce is exposed to the chlorine solution. Quick dips are much less effective than longer exposures. However, most of the sanitizing action of the chlorine is accomplished within the first several minutes of exposure. Prolonged exposure to strong chlorine solutions has been known to cause surface bleaching. Experience is the best guide to the correct combination of treatment time and chlorine concentration for the crop being processed.

Growth Stage of the Pathogen

Disease organisms may be either in the active vegetative form or in the form of spores. Chlorine will readily kill the vegetative form, but fungal spores are 10 to 1,000 times more difficult to kill. Therefore, chlorine treatment rarely eliminates all pathogens and sterilizes the surface of the produce. Many spores may remain on the surface to develop later should the opportunity arise. Further, chlorine kills only on contact, not systemically, and is effective only on exposed pathogens such as those suspended in water or those on the surface of produce; chlorine does not kill pathogens below the skin because it cannot contact them. Chlorination leaves no residual effect. Therefore, produce exposed to pathogens after treatment is susceptible to reinfection.

Wastewater Disposal

Chlorinated wastewater is customarily drained at the end of each work day or more often if circumstances dictate. This wastewater often contains sediment, pesticides, and other suspended matter. If it is discharged to a municipal wastewater treatment plant or to surface waters (canals, creeks, or ponds), regulatory agencies may consider it to be industrial wastewater. Land application of this material is normally permitted, but a nondischarge permit may be required. Operators may be required to obtain wastewater discharge permits. If you plan to use chlorination, check with the local office of the North Carolina Department of Environment, Health, and Natural Resources to determine whether a permit is required. Illegal disposal of hydrocooler wastewater may result in a substantial penalty.

The Environmental Protection Agency (EPA) regulation of May 8, 1991, (40 CFR part 180) exempts calcium hypochlorite and chlorine gas from residue tolerance requirements when it is used before or after harvest on any raw agricultural commodity. Any amount can legally be used; however, excessive chlorination can damage equipment, injure the surface of fruits and vegetables, and waste money. In addition, the use of excessive amounts of chlorine may pose a worker health and safety hazard that is regulated by the North Carolina Department of Labor.

Practical Rules for Successful Chlorination

- If water is not necessary in the packing process, do not use it: Wetting the produce greatly increases the likelihood of damage by postharvest diseases. If the produce must be washed to remove soil, there is no alternative to wetting. Hydrocooling also necessitates wetting the produce, although other methods, such as forced-air cooling, may be a viable option in some cases. When water is necessary in packing lines (for example, in dumping tanks, flumes, or a hydrocooler), always treat it to reduce the risk of disease.
- Monitor the chlorine concentration and the condition of the water: Check the chlorine concentration and pH frequently using test papers or electronic

equipment. Automatic chlorination equipment is available that will continually monitor the condition of the solution, add chlorine, and correct the pH. Also, monitor the water temperature.

- Avoid overexposure: Do not allow the produce to remain in contact with the solution longer than necessary. Check circulation patterns in chlorination tanks to eliminate dead spots.

- Change the water frequently: Chlorination efficiency is poor in dirty water. If necessary, wash very dirty produce with clean water before it comes into contact with the chlorinated water.

- Dispose of wastewater properly: Before installing chlorination equipment, plan how you will dispose of the wastewater. Land application of wastewater is normally allowed, but check to see if a permit is needed. Illegal disposal of chlorinated water could result in a substantial fine.

- Practice good sanitation: Hose off the packing equipment and floors daily; remove any dirt and trash that has settled in the chlorination tank. Sanitize the equipment with a spray solution composed of four pints of 5.25 percent sodium hypochlorite solution in 10 gallons of water. As an alternative, steam clean the equipment with an approved detergent. Do not allow culls or decayed produce to remain in or around the packing house.

- Protect workers: For their safety and comfort, workers must be protected from the chlorine fumes associated with excessively high levels of chlorine. If the amount of chlorine gas in the work area is great enough to cause worker discomfort, the amount of chlorine being used is well above that required for proper postharvest sanitation. If air monitoring equipment is not available, chlorine concentrations can be checked by asking a person who has not been desensitized by the odor to enter the work area. If he or she can smell the chlorine, the level is probably adequate. The concentration is too high if workers are continually irritated by the odor.

- Remember that chlorination will not solve all your problems: Even the best chlorination program may not be sufficient to prevent all postharvest decay. Prompt handling, proper sanitation, and rapid cooling should all be part of your postharvest disease management program. Produce infected in the field or otherwise damaged cannot be saved by chlorination.

References

- Post-harvest-diseases-meaning-classification-types-and-control-46346, plant-pathology: biologydiscussion.com, Retrieved 7 January, 2020
- Chlorination-and-postharvest-disease-control: content.ces.ncsu.edu, Retrieved 18 March, 2020

- Dubey, S.R. and Jalal, A.S. 2013. Species and Variety Detection of Fruits and Vegetables from Images. International Journal of Applied Pattern Recognition, 1(1), 108-126
- Main-types-of-penetration-of-post-harvest-pathogens-plant-diseases-43198, plant-diseases: biologydiscussion.com, Retrieved 9 June, 2020

6

Postharvest Packaging

Postharvest packaging is an important aspect of postharvest management. Temperature management is essential for packaging crops. All the diverse aspects of postharvest packaging such as packaging requirements, the different types of packaging and modified atmosphere packaging have been carefully analyzed in this chapter.

Packaging is done for more efficient handling and marketing, greater appeal, more potential life. Packaging requirement vary with different fruits and vegetables. Packaging cannot improve quality. Hence only best possible produce should be packed. Inclusion of decayed or damaged produced in bulk or consumer packages may become a source of infection and reduce the sale at the market. Packaging is not a substitute for refrigeration; packaging combined with refrigeration is the best methods. A good package is aim to protection of product from physical. Physiological and pathological deterioration causes throughout storage, transport and market.

Benefits of packaging serves as an efficient handling units, serves as a good storage unit, protect quality and reduce waste, protect from mechanical damage and moisture loss, provide beneficial modified atmosphere, prevent pilferage, provides service and sales motivation.

Material for Packaging

Wooden boxes, bamboo basket are the conventional packs. Fibre board cartoons, corrugated card boards, and several flexible plastic packaging. Materials used for packing material are: (1) Polyethylene (low density)-most widely used for consumer packaging,strong considerably moisture proof, resistant to several chemicals and cheap (2) Polyprophlene (3) Polyvinyl chloride film (4) Cellophane (5) Polifilm. The emphasis is being made now to use those materials which contains less or no wood for packaging as our forest resources being exhausted fastly and is at a precarious condition.

Consumer Packaging with Plastic

The original function of packaging was to contain carry and dispence products. However the use of plastic as packaging materials has allowed so much variation and versatility as to protect, presence, process, store, measure, communicate, and display of products. Fruits and vegetables are packaged in smaller quantities in polyethylene pouches as consumer packages. In these films proper ventilation is needed to prevent moisture accumulation. Leading to rotting of the content and to relegate O_2 and CO_2 concentration inside pack. High CO_2 concentration may cause deterioration in quality of the content. Congenial modified atmosphere inside the pack would increases the storage life of the contents.

Pre Packaging

- Pre-packaging increases the shelf life by creating modified atmosphere with an increase in concentration of CO_2 in package. L.D.P.E. films have high O_2 and CO_2 transmission rates are more durable.
- The pouch used reducing bruising facilitates inspection, reduces moisture losses and prevent dehydration. In pre-packaging leaves stalk stem are washed cleaned and weight quantities are put in pouches.
- Ethylene absorbents hydrate lime may insert in packages to retard ripening process.
- A wide range of packages like gunny bags, bamboo, woven, grass stem basket, wooden cares, earthen pots, corrugated fibre board cartoons and rigid plastic carats are used.
- Wheat and paddy straws, banana leaves, dry grass are used as cushioning material.

TYPES OF PACKAGES

There are as many types of packages available as there are products to put in them. Packaging systems are available in a variety of materials such as plastic, corrugated fiberboard, wood, and even sustainable materials such as bioplastics and fibers that decompose. One of the most common plastic packaging containers is the clear clamshell, manufactured from polyethylene terephthalate (PET) and other plastics using mechanical or vacuum thermoforming. Although plastic containers are necessary for certain commodities, corrugated and non-corrugated fiberboard is the dominant material used in fresh produce packaging. Wooden containers, usually wirebound, are a traditional form of produce packaging. They are an option for growers, although their use has gradually diminished over time because they are relatively heavy, expensive, and abrasive to the fruits and vegetables, and because they can present disposal issues. Sustainable packaging options are becoming increasingly more common and offer many advantages over traditional packaging containers. While beneficial to some, they are not appropriate for every operation. While there are a variety of functional packaging options available to growers of fresh fruits and vegetables, it is important to select the appropriate format for each specific commodity. Regardless of the material used, for a given commodity, it is important to use standard packaging sizes during the postharvest process so that growers can readily calculate total harvest by weight, count, and volume and thus more easily communicate production volumes to their buyers. Also, some buyers require that packaging footprints conform to the dimensions of the standard grocery pallet, which measures 40 x 48 inches (101.60 x 121.92 cm).

Packaging plays an important role in the fruit and vegetable distribution chain. According to Watkins and Nock, packaging serves four main functions:

- Containment: Containment is the basic requirement for movement of a product from one point to another. The package type and size will be a function of the product and market requirements.
- Protection and Preservation: Packages provide protection for the product against environmental factors such as dust and water, as well as impact and compression bruising, and friction injuries that can occur during handling and transport.
- Convenience: Products are packaged in sizes convenient for handlers and for the consumer. Consumer packages are often contained within larger containers for transport because of economies of scale. The product may be removed from the container and placed in a display as single consumer units, or repackaged for sale.
- Communication: In addition to advertising the type and source of the product, the package lists gross and net package weight, unit size of the product, and any additional information required by government regulations.

Another important feature of packaging for produce is that it should be designed with venting that facilitates cooling and temperature management during storage and transport so that field heat and respiratory heat can be efficiently removed.

Bulk Bins

Bulk bins allow producers to store, handle, and transport a large amount of product with relative ease. Bulk bins are manufactured from plastic polymers, wood, or corrugated fiberboard. Plastic and wood bulk bins are often only used on the farm because they are quite heavy and bulky. In terms of transportation, this restricts the amount of product available for shipping because each state imposes weight limits on transported goods. All types of bulk bins require heavy lifting equipment, such as forklifts, to move the product from field to packinghouse or storage because the contents often weigh in excess of 1,000 pounds.

Reusable plastic bulk bins (left) and wooden bulk bins (right) are used mostly on farms and at packinghouses.

Despite their weight, bulk bins have many advantages. They allow for easy movement of large quantities of product and are usually stackable. The ability to stack product is oftentimes essential for operations that are limited by the size of their packinghouse or floor space. Bulk bins also eliminate the waste that results from smaller, compartmentalized packaging containers that are single use and require more material to contain the same amount of product. Bulk bins produced from corrugated fiberboard are typically used by wholesalers, retailers, or other repacking operations because they are lightweight, may be collapsible, and even sometimes reusable. Corrugated fiberboard bulk bins can also have graphics on the side walls that can be customized for a specific commodity or with producer information.

Corrugated fiberboard bulk bins are sometimes used to ship less perishable crops like potatoes or, as shown here, watermelons.

For safety and sanitation, plastic bulk bins are preferred to wooden bulk bins and containers. The plastic surfaces are easier to clean, which should be done after every use. Because of their porous surface, wooden bulk bins and containers are very difficult, if not impossible, to clean and sanitize. If using a wooden container to transport product from the field to the packinghouse, you should wash the product after it has been removed from the wooden bulk bin. Alternatively, plastic liners may be used with wooden bins—both to reduce potential contamination and to minimize abrasion damage to the product from rough wood surfaces. Fasteners such as screws, nails, and bolts can also result in punctured and damaged product, requiring regular inspection of wooden bins. Corrugated fiberboard bins can be used more than once, but are more susceptible to damage than plastic and wooden bins. If reused, these bins should be inspected for cleanliness and covered in a plastic liner after the initial use in order to reduce the risk of cross-contamination. Alternatively, if the corrugated fiberboard bin has structural, water, or microbial damage, discard it immediately and use a new container in its place.

When determining whether to use bulk bins, consider the size and scale of an operation, the cost associated with the purchase and maintenance of the bins, as well as the commodity being packed. Smaller operations may have little use for large bulky bins or they may pack product directly into retail containers. Bulk bins, specifically those produced from plastic and wood, may be too expensive for some smaller producers. For commodities that are highly susceptible to bruising and mechanical damage during transportation and handling, the use of bulk bins would be inappropriate. However, other commodities, such as potatoes and watermelon, would be considered ideal candidates for bulk bins because they are able to resist most of the damage that occurs throughout the distribution chain. Considering these choices will help producers determine the best course of action for their specific operational needs.

Wooden Wirebound Containers

Throughout history, wood has been used for packaging, transporting, and storing a variety of products, including fresh produce. Because they are porous and absorbent, many believe that wood containers are not hygienic and therefore not suitable for fresh food products. However, it has been shown that certain species of wood, such as oak and pine, exhibit excellent antibacterial properties. While wooden wirebound containers will not completely eliminate the risk of product contamination from the spread of microorganisms, if properly maintained, they provide a cost-effective shortterm option for packing some commodities. Therefore, wood containers are a viable alternative for packaging, transporting, and storing fresh produce.

Wood containers come with or without handles and in a variety of sizes, shapes, and colors. Wood containers are available in several forms: barrels, baskets, hampers, or trays. Depending on the container, there may be a weave or specific design that can facilitate airflow for ventilation and rapid cooling. One of the most common forms of wood containers are the 1 1/9-bushel wirebound containers. This type of container is commonly used for commodities such as beans, sweetcorn, and potatoes.

Wood containers have many advantages that make them an excellent material for fresh produce. Wood containers are sturdy and thus able to endure mechanical stress and support heavier products in stacked loads. Wood containers are reusable if properly handled, maintained, cleaned, and sanitized, providing significant savings for the growers. Compared to fiberboard, wood containers are a better option for products that come into contact with water because they are more resistant to moisture. However, dry wood can absorb moisture from produce, causing items to shrivel when they come into contact with the dry wood. Compared to plastic, wood containers are more affordable and result in significant savings from that of reusable plastic containers (RPCs). Wooden wirebound containers should be packed correctly by an experienced worker; otherwise, overpacking may cause bulging and present problems for transporting. Unfortunately, wood will eventually deteriorate and decay and must be replaced. Depending on the commodity, a wood container may be a good alternative for small farmers who sell locally and can easily retrieve used containers, assuming that the distribution environment, application, and cost factors are also favorable.

Wooden wirebound produce containers (11/9 bushel).

Corrugated Fiberboard Cartons

Corrugated fiberboard cartons are comprised of a paperbased material manufactured from the Kraft paper pulping process. Corrugated fiberboard cartons are commonly made of single wall design with an inside and outside liner surrounding a fluted medium. The size of the flute determines the thickness of the corrugated fiberboard carton. Double wall and triple wall cartons are sometimes used for heavier and bulkier products because they provide added strength and rigidity. Traditionally, the strength of corrugated fiberboard cartons was measured using a burst test. However, burst tests measure the paper tensile strength whereas the stiffness and compression are typically the more desirable characteristics of corrugated fiberboard cartons. As a result, edge crush compression tests offer more valuable means of determining the performance of a corrugated fiberboard carton under compressive loads, which is important for containers holding products like fruits and vegetables that are commonly stacked.

Some produce is packaged into fiberboard cartons fitted with anti-moisture barriers to prevent the liquids from penetrating the fiberboard. As a result, corrugated fiberboard cartons with anti-moisture barriers are ideal for products that may be exposed to water at some point during postharvest handling. Traditionally, wax has been the primary substance applied to the fiberboard to prevent moisture from either entering or leaving the carton. However, it is rather difficult to recycle corrugated fiberboard that is saturated with wax. RPCs are an alternative method of packaging for a variety of fruits and vegetables, especially those that are commonly exposed to water.

Corrugated fiberboard cartons for tomatoes.

Corrugated fiberboard varies in flute size, burst strength, and edge crush strength, as well as surface treating and coatings. A few examples of products that are packaged into corrugated fiberboard include tomatoes, cucumbers, and sweet peppers; corrugated fiberboard is also used for the master containers of products like strawberries and grape tomatoes that are commonly packed first into plastic clamshells.

Custom-ordered corrugated fiberboard cartons may be too expensive for some producers. However, generic versions are available for lower cost. Some small farmers re-use cartons from larger operations, dismantling them, turning them inside-out to hide the original company's identity, and reconstructing them. This practice is not advised because the corrugated fiberboard cartons are prone to decay and there are food safety risks related to the crosscontamination of microbes carried on the re-used cartons.

Reusable Plastic Containers (RPCs)

Reusable plastic containers are becoming more prevalent in the fresh produce distribution system because producers and companies are trying to minimize their carbon footprint while simultaneously reducing costs. For many, RPCs have replaced corrugated fiberboard cartons as the preferred container for shipping fresh produce. Reusable plastic containers come in a wide variety of shapes and sizes, chosen primarily for their application to a specific commodity. Products that are hydrohandled or hydrocooled are especially suitable for packing in RPCs because RPCs are unaffected by moisture exposure. Reusable plastic containers are more durable and more rigid than their corrugated fiberboard counterparts. They are also weather resistant. As the name implies, RPCs can be utilized over and over again with very little wear, compared to most corrugated fiberboard cartons, which are used only once and then discarded. Most RPCs fold flat to minimize required space for storage and transport. Reusable plastic containers are produced from recyclable polymers, such as polypropylene (PP) or high density polyethylene (HDPE), and therefore can be melted and reformed into new products.

Zucchini summer squash in reusable plastic containers (RPCs).

RPCs can either be purchased or leased. Depending on the frequency of use, purchased RPCs can provide a significant cost savings over single-use corrugated fiberboard or wooden containers. While the initial cost of purchase and maintenance may seem too expensive, frequent use and economies of scale can allow for lower, long-term savings. Decide whether to use RPCS by considering the size and scale of the farming, packing, or shipping operation and the types of commodities with which the operation deals. Produce handlers must also have either closed distribution systems or return agreements with their receivers in order to utilize purchased RPCs. Otherwise, it will not be possible to recover the RPCs after they have been used (i.e., shipped).

RPCs must be cleaned between each use because residues from previous crops can lead to contamination of new harvests. Adhering to strict sanitization and cleaning procedures is necessary in order to reduce the risk of contamination when RPCs are reused. Chemicals, such as detergents, acids, or alkalis, can be used to remove dust, dirt, product residues, and various other debris items from the surface. Scrubbing while also using chemicals increases the probability of eliminating any contaminants that may lead to foodborne illnesses. Note that cleaning procedures alone cannot guarantee the complete removal of all microorganisms. In order to reduce or eliminate microorganisms, the RPCs must also be treated with sanitizers or disinfectants. Sanitizing agents, such as chlorine-based compounds, iodine compounds, or ammonium compounds, are most often applied to food contact surfaces to destroy or reduce the number of microorganisms.

While RPCs can provide significant savings to some producers, not all operations justify their utilization. The choice to purchase RPCs is dependent on the ability of the handler to maintain and clean those containers. Some producers may have the equipment and capacity to maintain and clean RPCs, while others may not. If a distribution system does not allow a handler's RPCs to be easily recovered, or if cleaning, sanitizing, and maintaining purchased RPCs is not feasible for an operation, then leasing is the better option. Leased RPCs are returned to the supplier by the receiver; the supplier is then responsible for cleaning and sanitizing the containers before reintroducing them into the distribution system. Producers should conduct a thorough cost analysis that considers customer (receiver) buy-in, material requirements, labor, transportation, inspections, maintenance, and cleaning before purchasing RPCs.

Insert Trays

Insert trays are used for a variety of reasons when packing fresh produce. Insert trays are typically formed from molded plastic or pulp to create an insert for a master container that has multiple cavities to contain individual fruit. The cavity is designed to accommodate a specific size and/or variety of commodity as well as to maximize the number of fruit per tray. Insert trays are relatively inexpensive, costing only pennies per tray and providing additional protection to the product to minimize damage.

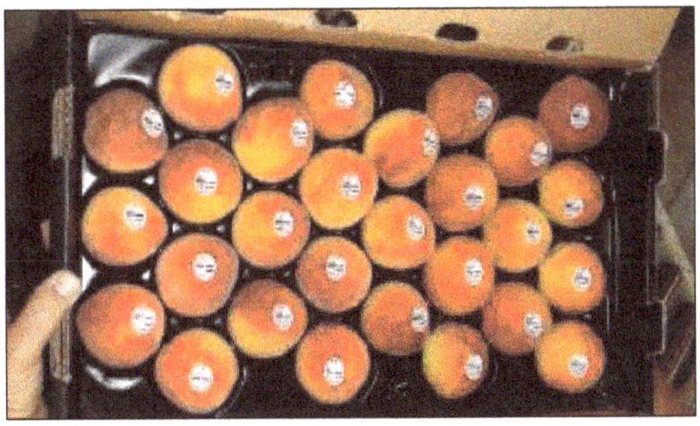

Thermoformed 28-count produce insert tray for peaches.

Plastic insert trays are usually thermoformed by placing large, heated, pliable sheets of plastic into molds where they conform to a customer's specified dimensions. This process is very similar to clamshell thermoforming. However, the sheets of plastic thermoformed for insert trays are much thinner than those used for clamshells. Plastic insert trays are typically formed from polypropylene (PP). Foam trays manufactured from expanded polyethylene (EPE) and other polymers are common as well. Molded paper pulp insert trays are produced by combining water and recycled or reclaimed newsprint or corrugated paper to form a slurry. The slurry is then formed in a mold to produce a tray with cavities for the intended commodity.

Insert trays provide additional physical protection to the product, facilitate hand sizing of produce items in fieldpacking operations, and prevent fruit-to-fruit spread of decay. Plastic insert trays help minimize abrasion damage, localize decay to individual fruit, and reduce moisture loss. Conversely, paper pulp insert trays do not perform well when wet. On the other hand, moisture retention in plastic trays, as a result of condensation or improper drying before or during the time of packing, can create areas of free moisture that encourage decay. This is a problem when the product rests in water trapped in the bottom of the plastic tray. Whether plastic or paper pulp, each tray is often used only once and then discarded. Therefore, the risk of food contamination is relatively low.

While plastic and pulp trays both serve to separate the products and keep them from coming into contact with one another, paper pulp and plastic foam insert trays have an additional benefit. Molded-pulp and foam insert trays provide more cushioning and support than plastic sheet insert trays. By absorbing shocks, the molded-pulp and foam insert trays protect the product from bruising during transportation and handling, therefore further minimizing postharvest losses. Both plastic sheet and foam insert trays exhibit disposal issues. Paper pulp insert trays are oftentimes produced from recycled materials, making them a more eco-friendly choice. However, if a producer's distribution system allows retrieval, plastic insert trays can be cleaned, sanitized and re-used. Overall, insert trays are an attractive option for packers and producers because they are low cost, hygienic, can be environmentally friendly, and result in fewer postharvest losses.

Clamshells

Clamshell containers are used extensively as consumer packages for a wide variety of commodities such as strawberries, blueberries, kumquats, and cherry tomatoes. A clamshell is a one-piece container consisting of two halves, forming the top and bottom, joined by a hinge with interlocking fasteners or snaps designed for closure. Clamshell containers are manufactured from a variety of different polymers such as polystyrene (PS) and polyvinyl chloride (PVC). However, one of the most common polymers used in clamshell containers is polyethylene terephthalate (PET) because it is highly transparent and easily recycled. Properly designed clamshells have a sufficient number of vent holes to facilitate cooling, to allow the product to "breathe" (i.e., allow exchange of respiratory heat and gases), and to allow for excess moisture to escape. The structure is often made using thermoforming or injection molding and can be customized into a variety of desired shapes and sizes.

Polyethylene terephthalate thermoformed (PET) clamshell containers for strawberries.

Clamshell containers also provide an area for attaching labels that include required information such as nutrition, weight, and manufacturer identification in addition to marketing and consumer information. Some of the advantages of clamshell containers include complete product visibility, product security, and increased perceived value. The container also prevents the contents from being handled by multiple individuals throughout the distribution chain, providing an extra degree of food safety. Although they are recyclable, clamshell containers are produced from non-renewable fossil fuels. Therefore, the polymers used to manufacture the containers are not the most environmentally friendly. However, clamshells made from biodegradable biopolymers such as polylactic acid (PLA), which is derived from corn starch or sugar cane dextrose, are available.

Choosing clamshell containers as the preferred method of packaging is often dependent on the commodity, industry standards, and consumer demand. Historically, smaller, open-air containers were used before clamshells. However, clamshells are the preferred choice for producers and packers alike.

Clamshells may be too expensive, however, for small farm operations. An alternative to clamshells might be perforated plastic bags with hand-punched holes for ventilation. Although bags do not provide protection from physical injuries like clamshells, they do protect produce from moisture loss and abrasion. A master container of some sort would also be required to hold consumer-size bags during handling between the farm and point-of-sale. However, perforated plastic bags would not be acceptable for many types of products grown by small farmers. For example, with tomatoes and peppers, the weight of the product would limit use of perforated plastic bags because the bags may stretch, damaging their integrity, and ultimately breaking. However, perforated plastic bags would be an ideal choice for leafy greens, grapes, or radishes.

Sustainable Packaging

Sustainable produce packaging is becoming increasingly more common in the marketplace, driven, in part, by consumer demand and government regulation for environmentally friendly materials. While fresh produce is healthy and good for human consumption, many packaging options used to protect fresh products do not contain environmentally friendly materials. Much of that packaging is used only once and then discarded, never to be recycled or reused.

Broadly, sustainable packaging is intended to be functional, but it should also be cost effective. A drastic increase in costs, relative to traditional packaging options, may render it too expensive for use. Sustainable packaging may also fulfill more specific objectives in its design and functionality. For example, reducing or minimizing materials will reduce the weight or volume that the packaging occupies while maintaining logistics efficiency throughout the products' life cycle in the distribution chain will reduce waste. Logistics efficiency involves efficient transportation, design space utilization, and reduced tare weight. The use of biodegradable materials or reusable packaging can minimize the amount of packaging destined for landfills or recycling streams, ultimately resulting in a lower amount of accumulated waste.

There are many reasons why producers choose sustainable packaging options. Economic, environmental, or even ethical concerns can justify choosing sustainable packaging over other options. Ultimately, each producer must weigh the costs and benefits of adopting sustainable packaging formats in order to determine which option is most appropriate for them.

PACKAGING REQUIREMENTS

Packaging fresh fruits and vegetables is one of the more important steps in the long and complicated journey from grower to consumer. Bags, crates, hampers, baskets, cartons, bulk bins, and palletized containers are convenient containers for handling, transporting, and marketing fresh produce. More than 1,500 different types of packages are used for produce in the United States and the number continues to increase as the industry introduces new packaging materials and concepts. Although the industry generally agrees that container standardization is one way to reduce cost, the trend in recent years has moved toward a wider range of package sizes to accommodate the diverse needs of wholesalers, consumers, food service buyers, and processing operations.

Packing and packaging materials contribute a significant cost to the produce industry; therefore it is important that packers, shippers, buyers, and consumers have a clear understanding of the wide range of packaging options available.

Function of Packaging

A significant percentage of produce buyer and consumer complaints may be traced to container failure because of poor design or inappropriate selection and use. A properly designed produce container should contain, protect, and identify the produce, satisfying everyone from grower to consumer.

Packaging Points

- Recyclability/Biodegradability: A growing number of US markets and many export markets have waste disposal restrictions for packaging materials. In the near future, almost all produce packaging will be recyclable or biodegradable, or both. Many of the largest buyers of fresh produce are also those most concerned about environmental issues.
- Variety: The trend is toward greater use of bulk packages for processors and wholesale buyers and smaller packages for consumers. There are now more than 1,500 different sizes and styles of produce packages.
- Sales Appeal: High quality graphics are increasingly being used to boost sales appeal. Multi-color printing, distinctive lettering, and logos are now common.
- Shelf Life: Modern produce packaging can be custom engineered for each commodity to extend shelf life and reduce waste.

Containment

The container must enclose the produce in convenient units for handling and distribution. The produce should fit well inside the container, with little wasted space. Small produce items that are spherical or oblong (such as potatoes, onions, and apples) may be packaged efficiently utilizing a variety of different package shapes and sizes. However, many produce items such as asparagus, berries, or soft fruit may require containers specially designed for that item. Packages of produce commonly handled by hand are usually limited to 50 pounds. Bulk packages moved by forklifts may weigh as much as 1,200 pounds.

Protection

The package must protect the produce from mechanical damage and poor environmental conditions during handling and distribution. To produce buyers, torn, dented, or collapsed produce packages usually indicate lack of care in handling the contents. Produce containers must be sturdy enough to resist damage during packaging, storage, and transportation to market. Because almost all produce packages are palletized, produce containers should have sufficient stacking strength to resist crushing in a low temperature, high humidity environment. Although the cost of packaging materials has escalated sharply in recent years, poor quality, lightweight containers that are easily damaged by handling or moisture are no longer tolerated by packers or buyers.

Produce destined for export markets requires containers to be extra sturdy. Air-freighted produce may require special packing, package sizes, and insulation. Marketers who export fresh produce should consult with freight companies about any special packaging requirements. Additionally, the USDA and various state export agencies may be able to provide specific packaging information.

Damage resulting from poor environmental control during handling and transit is one of the leading causes of rejected produce and low buyer and consumer satisfaction. Each fresh fruit and vegetable commodity has its own requirements for temperature, humidity, and environmental gas composition. Produce containers should be produce friendly - helping to maintain an optimum environment for the longest shelf life. This may include special materials to slow the loss of water from the produce, insulation materials to keep out the heat, or engineered plastic liners that maintain a favorable mix of oxygen and carbon dioxide.

Identification

The package must identify and provide useful information about the produce. It is customary (and may be required in some cases) to provide information such as the produce name, brand, size, grade, variety, net weight, count, grower, shipper, and country of origin. It is also becoming more common to find included on the package nutritional information, recipes, and other useful information directed specifically at the consumer. In consumer marketing, package appearance has also become an important part of point of sale displays.

Universal Product Codes (UPC or bar codes) may be included as part of the labeling. The UPCs used in the food industry consist of a ten-digit machine readable code. The first five digits are a number assigned to the specific producer (packer or shipper) and the second five digits represent specific product information such as type of produce and size of package. Although no price information is included, UPCs are used more and more by packers, shippers, buyers, and retailers as a fast and convenient method of inventory control and cost accounting. Efficient use of UPCs requires coordination with everyone who handles the package.

Types of Packaging Materials

Wood

Pallets literally form the base on which most fresh produce is delivered to the consumer. Pallets were first used during World War II as an efficient way to move goods. The produce industry uses approximately 190 of the 700 million pallets produced per year in the United States. About 40 percent of these are single-use pallets. Because many are of a non-standard size, the pallets are built as inexpensively as possible and discarded after a single use. Although standardization efforts have been slowly under way for many years, the efforts have been accelerated by pressure from environmental groups, in addition to the rising cost of pallets and landfill tipping fees.

Over the years, the 40-inch wide, by 48-inch long pallet has evolved as the unofficial standard size. Standardization encourages re-use, which has many benefits. Besides reducing cost because they may be used many times, most pallet racks and automated pallet handling equipment are designed for standard-size pallets. Standard size pallets make efficient use of truck and van space and can accommodate heavier loads and more stress than lighter single-use pallets. Additionally, the use of a single pallet size could substantially reduce pallet inventory and warehousing costs along with pallet repair and disposal costs. The adoption of a pallet standard throughout the produce industry would also aid efforts toward standardization of produce containers.

In the early 1950s, an alternative to the pallet was introduced. It is a pallet-size sheet (slipsheet) of corrugated fiberboard or plastic (or a combination of these materials) with a narrow lip along one or more sides. Packages of produce are stacked directly on this sheet as if it were a pallet. Once the packages are in place, they are moved by a specially equipped forklift equipped with a thin metal sheet instead of forks.

Slipsheets are considerably less expensive than pallets to buy, store, and maintain; they may be re-used many times; and they reduce the tare weight of the load. However, they require the use of a special forklift attachment at each handling point from packer to retailer.

Depending on the size of produce package, a single pallet may carry from 20 to over 100 individual packages. Because these packages are often loosely stacked to allow for air circulation, or are bulging and difficult to stack evenly, they must be secured (unitized) to prevent shifting during handling and transit. Although widely used, plastic straps and tapes may not have completely satisfactory results. Plastic or paper corner tabs should always be used to prevent the straps from crushing the corners of packages.

Plastic stretch film is also widely used to secure produce packages. A good film must stretch, retain its elasticity, and cling to the packages. Plastic film may conform easily to various size loads. It helps protect the packages from loss of moisture, makes the pallet more secure against pilferage, and can be applied using partial automation. However, plastic film severely restricts proper ventilation. A common alternative to stretch film is plastic netting, which is much better for stabilizing some pallet loads, such as those that require forced-air cooling. Used stretch film and plastic netting may be difficult to properly handle and recycle.

A very low-cost and almost fully automated method of pallet stabilization is the application of a small amount of special glue to the top of each package. As the packages are stacked, the glue secures all cartons together. This glue has a low tensile strength so cartons may be easily separated or repositioned, but a high shear strength so they will not slide. The glue does not present disposal or recycling problems.

Pallet Bins

Substantial wooden pallet bins of milled lumber or plywood are primarily used to move produce from the field or orchard to the packing house. Depending on the application, capacities may range from 12 to more than 50 bushels. Although the height may vary, the length and width is generally the same as a standard pallet (48 inches by 40 inches). More efficient double-wide pallet bins (48 inches by 80 inches) are becoming more common in some produce operations.

Most pallet bins are locally made; therefore it is very important that they be consistent from lot to lot in materials, construction, and especially size. For example, small differences in overall dimensions can add up to big problems when several hundred are stacked together for cooling, ventilation, or storage. It is also important that stress points be adequately reinforced. The average life of a hardwood pallet bin that is stored outside is approximately five years. When properly protected from the weather, pallets bins may have a useful life of 10 years or more.

Wire-bound Crates

Although alternatives are available, wooden wire-bound crates are used extensively for snap beans, sweet corn and several other commodities that require hydrocooling. Wire-bound crates are sturdy, rigid and have very high stacking strength that is essentially unaffected by water. Wire-bound crates come in many different sizes from half-bushel to pallet-bin size and have a great deal of open space to facilitate cooling and ventilation. Although few are re-used, wire-bound crates may be dissembled after use and shipped back to the packer (flat). In some areas, used containers may pose a significant disposal problem. Wire-bound crates are not generally acceptable for consumer packaging because of the difficulty in affixing suitable labels.

Wooden Crates and Lugs

Wooden crates, once extensively used for apples, stone fruit, and potatoes have been almost totally replaced by other types of containers. The relative expense of the container, a greater concern for tare weight, and advances in material handling have reduced their use to a few specialty items, such as expensive tropical fruit. The 15-, 20-, and 25-pound wooden lugs still used for bunch grapes and some specialty crops are being gradually replaced with less costly alternatives.

Wooden Baskets and Hampers

Wire-reinforced wood veneer baskets and hampers of different sizes were once used for a wide variety of crops from strawberries to sweetpotatoes. They are durable and may be nested for efficient transport when empty. However, cost, disposal problems, and difficulty in efficient palletization have severely limited their use to mostly local grower markets where they may be re-used many times.

Corrugated Fiberboard

Corrugated fiberboard (often mistakenly called cardboard or pasteboard) is manufactured in many different styles and weights. Because of its relativity low cost and versatility, it is the dominant produce container material and will probably remain so in the near future. The strength and serviceability of corrugated fiberboard have been improving in recent years.

Most corrugated fiberboard is made from three or more layers of paperboard manufactured by the kraft process. To be considered paperboard, the paper must be thicker than 0.008 inches. The grades of paperboard are differentiated by their weight (in pounds per 1,000 square feet) and their thickness. Kraft paper made from unbleached pulp has a characteristic brown color and is exceptionally strong. In addition to virgin wood fibers, Kraft paper may have some portion of synthetic fibers for additional strength, sizing (starch), and other materials to give it wet strength and printability. Most fiberboard contains some recycled fibers. Minimum amounts of recycled materials may be specified by law and the percentage is expected to increase in the future. Tests have shown that cartons of fully recycled pulp have about 75 percent of the stacking strength of virgin fiber containers. The use of recycled fibers will inevitably lead to the use of thicker walled containers.

Double-faced corrugated fiberboard is the predominant form used for produce containers. It is produced by sandwiching a layer of corrugated paperboard between an inner and outer liner (facing) of paper-board. The inner and outer liner may be identical, or the outer layer may be preprinted or coated to better accept printing. The inner layer may be given a special coating to resist moisture. Heavy-duty shipping containers, such as corrugated bulk bins that are required to have high stacking strength, may have double- or even triple-wall construction. Corrugated fiberboard manufacturers print box certificates on the bottom of containers to certify certain strength characteristics and limitations. There are two types of certification. The first certifies the minimum combined weight of both the inner and outer facings and that the corrugated fiberboard material is of a minimum bursting strength. The second certifies minimum edge crush test (ETC) strength. Edge crush strength is a much better predictor of stacking strength than is bursting strength. For this reason, users of corrugated fiberboard containers should insist on ECT certification to compare the stackability of various containers. Both certificates give a maximum size limit for the container (sum of length, width, and height) and the maximum gross weight of the contents.

Both cold temperatures and high humidities reduce the strength of fiberboard containers. Unless the container is specially treated, moisture absorbed from the surrounding air and the contents can reduce the strength of the container by as much as 75 percent. New anti-moisture coatings (both wax and plastic) are now available to substantially reduce the effects of moisture.

Waxed fiberboard cartons (the wax is about 20 percent of fiber weight) are used for many produce items that must be either hydrocooled or iced. The main objection to wax cartons is disposal after use— wax cartons cannot be recycled and are increasingly being refused at landfills. Several states and municipalities have recently taxed wax cartons or have instituted rigid back haul regulations. Industry sources suggest that wax cartons will eventually be replaced by plastic or, more likely, the use of ice and hydrocooling will be replaced by highly controlled forced-air cooling and rigid temperature and humidity maintenance on many commodities.

In many applications for corrugated fiberboard containers, the stacking strength of the container is a minor consideration. For example, canned goods carry the majority of their own weight when stacked. Fresh produce usually cannot carry much of the vertical load without some damage. Therefore, one of the primarily desired characteristics of corrugated fiberboard containers is stacking strength to protect the produce from crushing. Because of their geometry, most of the stacking strength of corrugated containers is carried by the corners. For this reason, hand holes and ventilation slots should never be positioned near the corners of produce containers and be limited to no more than 5 to 7 percent of the side area.

Interlocking the packages (cross stacking) is universally practiced to stabilize pallets. Cross stacking places the corner of one produce package at the middle of the one below it, thus reducing its stacking strength. To reduce the possibility of collapse, the first several layers of each pallet should be column stacked (one package directly above the other). The upper layers of packages may be cross stacked as usual with very little loss of pallet stability.

There are numerous styles of corrugated fiberboard containers. The two most used in the produce industry are the one piece, regular slotted container (RSC) and the two piece, full telescoping container (FTC). The RSC is the most popular because it is simple and economical. However, the RSC has relatively low stacking strength and therefore must be used with produce, such as potatoes, that can carry some of the stacking load. The FTC, actually one container inside another, is used when greater stacking strength and resistance to bulging is required. A third type of container is the Bliss box, which is constructed from three separate pieces of corrugated fiberboard. The Bliss box was developed to be used when maximum stacking strength is required. The bottoms and tops of all three types of containers may be closed by glue, staples, or interlocking slots.

Almost all corrugated fiberboard containers are shipped to the packer flat and assembled at the packing house. To conserve space, assembly is usually performed just before use. Assembly may be by hand, machine, or a combination of both. Ease of assembly should be carefully investigated when considering a particular style of package.

In recent years, large double-wall or even triple- wall corrugated fiberboard containers have increasingly been used as one-way pallet bins to ship bulk produce to processors and retailers. Cabbage, melons, potatoes, pumpkins, and citrus have all been shipped successfully in these containers. The container cost per pound of produce is as little as one fourth of traditional size containers. Some bulk containers may be collapsed and re-used.

For many years, labels were printed on heavy paper and glued or stapled to the produce package. The high cost of materials and labor has all but eliminated this practice. The ability to print the brand, size, and grade information directly on the container is one of the greatest benefits of corrugated fiberboard containers. There are basically two methods used to print corrugated fiberboard containers:

- Post Printed: When the liner is printed after the corrugated fiberboard has been formed, the process is known as post printing. Post printing is the most widely used printing method for corrugated fiberboard containers because it is economical and may be used for small press runs. However, postprinting produces graphics with less detail and is usually limited to one or two colors.
- Preprinted: High quality, full-color graphics may be obtained by preprinting the linerboard before it is attached to the corrugated paperboard. Whereas the cost is about 15 percent more than standard two color containers, the eye catching quality of the graphics makes it very useful for many situations. The visual quality of the package influences the perception of the product because the buyer's

first impression is of the outside of the package. Produce managers especially like high quality graphics that they can use in supermarket floor displays.

- Preprinted cartons are usually reserved for the introduction of new products or new brands. Market research has shown that exporters may benefit from sophisticated graphics. The increased cost usually does not justify use for mature products in a stable market, but this may change as the cost of these containers becomes more competitive.

Pulp Containers

Containers made from recycled paper pulp and a starch binder are mainly used for small consumer packages of fresh produce. Pulp containers are available in a large variety of shapes and sizes and are relatively inexpensive in standard sizes. Pulp containers can absorb surface moisture from the product, which is a benefit for small fruit and berries that are easily harmed by water. Pulp containers are also biodegradable, made from recycled materials, and recyclable.

Paper and Mesh Bags

Consumer packs of potatoes and onions are about the only produce items now packed in paper bags. The more sturdy mesh bag has much wider use. In addition to potatoes and onions, cabbage, turnips, citrus, and some specialty items are packed in mesh bags. Sweet corn may still be packaged in mesh bags in some markets. In addition to its low cost, mesh has the advantage of uninhibited air flow. Good ventilation is particularly beneficial to onions. Supermarket produce managers like small mesh bags because they make attractive displays that stimulate purchases.

However, bags of any type have several serious disadvantages. Large bags do not palletize well and small bags do not efficiently fill the space inside corrugated fiberboard containers. Bags do not offer protection from rough handling. Mesh bags provide little protection from light or contaminants. In addition, produce packed in bags is correctly perceived by the consumer to be less than the best grade. Few consumers are willing to pay premium price for bagged produce.

Plastic Bags

Plastic bags (polyethylene film) are the predominant material for fruit and vegetable consumer packaging. Besides the very low material costs, automated bagging machines further reduce packing costs. Film bags are clear, allowing for easy inspection of the contents, and readily accept high quality graphics. Plastic films are available in a wide range of thicknesses and grades and may be engineered to control the environmental gases inside the bag. The film material "breathes" at a rate necessary to maintain the correct mix of oxygen, carbon dioxide, and water vapor inside the bag. Since each produce item has its own unique requirement for environmental gases, modified atmosphere packaging material must be specially engineered for each item. Research has shown that the shelf life of fresh produce is extended considerably by this packaging. The explosive growth of precut produce is due in part to the availability of modified atmosphere packaging.

In addition to engineered plastic films, various patches and valves have been developed that affix to low-cost ordinary plastic film bags. These devices respond to temperature and control the mix of environmental gases.

Shrink Wrap

One of the newest trends in produce packaging is the shrink wrapping of individual produce items. Shrink wrapping has been used successfully to package potatoes, sweet-potatoes, apples, onions, sweet corn, cucumbers and a variety of tropical fruit. Shrink wrapping with an engineered plastic wrap can reduce shrinkage, protect the produce from disease, reduce mechanical damage and provide a good surface for stick-on labels.

Rigid Plastic Packages

Packages with a top and bottom that are heat formed from one or two pieces of plastic are known as clamshells. Clamshells are gaining in popularity because they are inexpensive, versatile, provide excellent protection to the produce, and present a very pleasing consumer package. Clamshells are most often used with consumer packs of high value produce items like small fruit, berries, mushrooms, etc., or items that are easily damaged by crushing. Clamshells are used extensively with precut produce and prepared salads. Molded polystyrene and corrugated polystyrene containers have been test marketed as a substitute for waxed corrugated fiberboard. At present they are not generally cost competitive, but as environmental pressures grow, they may be more common. Heavy-molded polystyrene pallet bins have been adopted by a number of growers as a substitute for wooden pallet bins. Although at present their cost is over double that of wooden bins, they have a longer service life, are easier to clean, are recyclable, do not decay when wet, do not harbor disease, and may be nested and made collapsible.

As environmental pressures continue to grow, the disposal and recyclability of packaging material of all kinds will become a very important issue. Common polyethylene may take from 200 to 400 years to breakdown in a landfill. The addition of 6 percent starch will reduce the time to 20 years or less. Packaging material companies are developing starch-based polyethylene substitutes that will break down in a landfill as fast as ordinary paper.

The move to biodegradable or recyclable plastic packaging materials may be driven by cost in the long term, but by legislation in the near term. Some authorities have proposed a total ban on plastics. In this case, the supermarket of the early 21st century may resemble the grocery markets of the early 20th century.

Standardization

Produce package standardization is interpreted differently by different groups. The wide variety of package sizes and material combinations is a result of the market responding to demands from many different segments of the produce industry. For example, many of the large-volume buyers of fresh produce are those most concerned with the environment. They demand less packaging and the use of more recyclable and biodegradable materials, yet would also like to have many different sizes of packages for convenience. Packers want to limit the variety of packages they must carry in stock, yet they have driven the trend toward preprinted, individualized containers. Shippers and trucking companies want to standardize sizes so the packages may be better palletized and handled.

Produce buyers are not a homogeneous group. Buyers for grocery chains have different needs than buyers for food service. For grocery items normally sold in bulk, processors want largest size packages that they can handle efficiently - to minimize unpacking time and reduce the cost of handling or disposing of the used containers. Produce managers, on the other hand, want individualized, high quality graphics to entice retail buyers with in-store displays.

Selecting the right container for fresh produce is seldom a matter of personal choice for the packer. For each commodity, the market has unofficial, but nevertheless rigid standards for packaging; therefore it is very risky to use a nonstandard package. Packaging technology, market acceptability, and disposal regulations are constantly changing. When choosing a package for fresh fruits and vegetables, packers must consult the market, and in some markets, standard packages may be required by law.

MODIFIED ATMOSPHERE PACKAGING

Modified atmosphere packaging (MAP) is a useful preservation system that allows to significantly increase the shelf life of fruits and vegetables. The MAP results of changing the composition of the atmosphere in the packaging headspace due to the dynamic interaction between the metabolic processes of the packaged product on the one hand, in which O_2 is consumed and another gases such as CO_2 and water vapor are generated, and on the other hand, by transferring all of these gases through the package. The aim of the system is to balance these two processes in such a way that constant levels of these different gases are reached in the packaging headspace and that these equilibrium levels are as favorable as possible to preserve the product.

After harvest, the fresh produce undergoes a series of biochemical transformations that lead to the development of color, aroma and characteristic flavors, reduction of acidity and tissue softening. These processes continue until the substrates that support them are consumed, after which follows the deterioration or senescence phase. The postharvest period involves a series of biochemical reactions or metabolic activities that can lead to an increase in the concentration of organic acids, lipids, phenolic compounds, generation of volatile compounds (aroma), variations in the activity of enzymes, degradation of chlorophyll and biosynthesis of pigments, degradation of pectins and conversion of starch in sugars, which will cause loss of firmness and weight, and increase in the sweetness perception.

In the case of edible products, all previous processes lead to the development of characteristics that make them acceptable for consumption. In this period, the plant cells continue their respiratory activity, using oxygen from the surrounding atmosphere and releasing carbon dioxide as result of a series of oxide-reduction reactions of the substrates present.

The respiratory response of fresh produce in postharvest and the development of ripening and senescence processes depends on the temperature, of the atmosphere surrounding the stored product and also on the levels of gases such as O_2, CO_2, ethylene and water vapor, which strongly influence its metabolism. For this reason, by regulating the temperature and gas concentrations in the atmosphere surrounding the product, it is possible to reduce the development of its different biochemical processes and increase its preservation time. Likewise, the growth of microorganisms that may cause deterioration or other potentially pathogenic organisms will also be closely related to environmental conditions due to their development, and activity can be controlled by modifying these conditions.

Modified atmosphere packaging (MAP) has been constituted as a preservation method that allows to significantly increase the shelf life of fruits and vegetables. The MAP results of changing the composition of the atmosphere surrounding the product in order to reduce its natural deterioration and the microbial spoilage.

In a MAP system, there is a change in the concentration of gases in the packaging headspace due to the dynamic interaction between the metabolic and biochemical processes of the packaged product on the one hand, in which O_2 is consumed and CO_2, ethylene and water vapor are generated, and on the other hand, the transfer of all of these gases through the packaging. Accordingly, O_2 from the external atmosphere will be entering through the packaging to replace that the product is consuming and the other gases will be leaving out through the packaging system as surplus. The aim of the system is to balance these two processes in such a way that constant levels of these different gases are reached in the packaging headspace and that these equilibrium levels are as favorable as possible to preserve the product.

In the first MAP systems, a reference was made to active and passive atmospheres depending on whether a volume of gas with a different atmospheric concentration was introduced at the time of sealing the package (active MAP), or simply the bag was sealed with atmospheric air (passive MAP). This is of limited utility if it is not considered that regardless of the initial concentration of gases in the packaging headspace, the final concentration of these will depend on the interaction between the metabolic processes of the product (and possible microorganisms present) and the exchange of gases through the packaging. Failure to know the different rates at which these processes are carried out will result in an inadequate MAP, resulting in concentrations of gases that do not contribute to preserving the product at the best or that directly increase its deterioration at the worst. For this reason and in order to obtain a satisfactory MAP, it is necessary to design the packaging system previously, and this basically consists of determining the gas transfer capacity in the packaging that is required to balance the metabolic processes of the product to be packed and in this way, to achieve favorable gas concentrations for its preservation.

After packing a fruit or vegetable, the O_2 concentration in the headspace will decrease, and the CO_2 concentration will increase until reaching constant values once the equilibrium is reached. In these conditions of reduced O_2 and high CO_2, the metabolic processes of the product and the microbial activity slow down increasing its shelf life. Likewise, as the storage temperature decreases, there is also a reduction in the development of these processes resulting in an additional increase of the produce shelf life. On the other hand, low O_2 levels, high CO_2 levels and above all, low temperatures in the packaging system can significantly decrease the growth of microorganisms that cause deterioration in the product and of other ones potentially pathogenic. In these unfavorable circumstances for microbial growth (low O_2, high CO_2 and low temperature), a competition is established between the produce, natural microflora and other external and potentially pathogenic microorganisms that has an inhibitory effect for the development of the latter ones. As shown in table, for a wide range of fresh produce, the most favorable levels in the packaging headspace for preservation are between 1 and 10% of O_2, 5 and 15% of CO_2 and 80 and 90% RH, with temperatures as low as possible (0–15 °C) but without inducing chilling injury.

Table: Suitable O_2 and CO_2 concentrations in MAP for different fruits and vegetables.

Produce	O_2 (%)	CO_2 (%)
Fruits		
Apple	2–3	1–3
Avocado	2–5	3–10
Banana	2–5	2–8
Feijoa	5–11	7–12
Mango	3–7	5–8
Orange	5–10	5–8
Pear	2–3	2–5
Strawberry	5–10	5–15
Vegetables		
Broccoli	2–3	5–10

Carrot	2–5	2–5
Lettuce	2–3	1–3
Onion	2–5	2–10
Pepper	3–5	1–3
Radish	2–5	2–5
Spinach	15–20	10–20
Tomato	3–5	5–11

MAP uses to Preserve Vegetables and Fruits

MAP has been successfully applied to preserve several vegetables including broccoli, cauliflower, carrots, garlic and others. The effect of the packaging has been studied on the quality of broccoli and cauliflowers. In order to preserve these products, micro or macro perforated polyethylene and polypropylene bags have been used to compensate for the respiration and transpiration of the packed product and to achieve favorable equilibrium gas levels for its preservation. Other vegetables have been successfully preserved in MAP systems, including cabbage, cucumber, onion, spinach and tomato. Refrigeration temperatures were used in all cases, although at room temperatures, the packaging system may continue to have a preservation effect, although somewhat lower.

As for fruits, MAP systems have been used successfully to preserve apple, banana, breba, blueberry, feijoa, grape, guava, mandarin, medlar, papaya, pear, pomegranate, strawberry and others. In all these cases, it was possible to increase the product shelf life between 40% and up to twice with respect to atmospheric conditions. Recommended O_2 and CO_2 concentrations in the MAP system are shown in Table for various fruits and vegetables.

Packaging Materials used in MAP

Most MAP systems consist of flexible or semiflexible structures, bags of different sizes and thicknesses with and without perforations. The most widely used polymer films for MAP are synthetic polyolefins, low-density polyethylene (LDPE), linear low-density polyethylene (LLDPE), high-density polyethylene (HDPE), polypropylene (PP), polyvinyl chloride (PVC), polyesters, polyethylene terephthalate (PET), polyvinylidene chloride (PVDC), ethylene-vinyl alcohol (EVOH), polyamide (Nylon), polyvinyl alcohol (PVOH), ethylene vinyl acetate (EVA), cellulose-derived plastics such as cellophane and natural biodegradable polymers like polylactic acid (PLA). Each of these polymeric materials offers different mechanical characteristics and fundamentally different permeability to O_2, CO_2 and water vapor. Depending on the specific requirements of gas permeation, materials of very low permeability to O_2 and CO_2 such as polyamide, of moderate permeability such as polypropylene and EVA, or of high permeability as LDPE can be selected or even combinations of these materials can be made in several layers to adjust the MAP system to the produce respiration and transpiration. In addition, in order to avoid the accumulation of moisture inside the package while maintaining a reduced permeability to O_2 and CO_2, it is possible to use PLA or EVOH, which have a high permeability to water vapor. In table, permeability (Q) coefficients to O_2, CO_2 and water vapor are for different polymeric materials.

Table: Permeability coefficients ($cm^3\ mm\ m^{-2}\ atm^{-1}\ d^{-1}$) at 20 °C for different packaging materials.

Polymeric material	Q_{O2}	Q_{CO2}	Q_{N2}	Q_{H2O}^{*}
Low-density Polyethylene (LDPE)	190–200	1050–1250	100–150	5500–6000
High-density Polyethylene (HDPE)	40–70	160–190	14–20	2100–2500
Cast Polypropylene (PP)	80–95	250–280	17–25	4000–4200
Oriented Polypropylene (OPP)	40–50	180–200	10–15	1800–2200
Polylactic Acid (PLA)	60–80	150–190	20–30	300000–350000
Ethylene vinyl acetate (EVA)	300–330	1100–1300	120–150	10000–15000
Polyamide (PA)	1–5	3–10	0.3–1	20000–30000

Dynamics of a MAP System

In the design of MAP systems for fresh produce, it is necessary to understand the dynamics of the interactions established between the product, the atmosphere generated in the packaging headspace and the packaging system itself. Each MAP design must be optimized for a specific product, since agricultural products have different metabolisms to each other and, the MAP system must balance the processes of respiration, transpiration and gas permeation through the packaging that will be occurring simultaneously. Factors that affect both the metabolism of the packaged product and the gas permeation through the packaging should be considered for the design of the MAP system. Among these factors are the type of produce and ripening stage, the characteristics of the packaging system and the storage conditions. With regard to the packaged product, it is also necessary to know in advance its transpiration, oxygen consumption and CO_2 and ethylene generation rates at the packaging conditions and the optimum concentrations of these gases and relative humidity that favor the increase of its shelf life.

For the proper design of the MAP system, different mathematical models have been developed to represent the processes of respiration, transpiration and ethylene production for a variety of horticultural products. In these models, it is intended to represent the rates of O_2 consumption and CO_2, ethylene and water vapor generation from the product as a function of the storage temperature, the concentration of these gases in the packaging headspace and the weight of the packaged product. In some studies, these metabolic processes developed in the fresh produce are described by using enzymatic kinetics equations, while in other studies, empirical or semiempirical equations are used to represent them. As for the transfer of gases through the packaging and the perforations that it may have, generally have been developed models derived from the

Fick's law, stating the mass balances for each gas in the system and considering the influence of the temperature at the gas permeation rates.

Mathematical Representation of the MAP System

The change in the composition of each gas in the packaging headspace will depend on the metabolic processes of the packaged product and the gas transfer through the packaging system.

$$\begin{bmatrix} \text{Accumulation or} \\ \text{decrease rate of gas in} \\ \text{the packaging headspace} \end{bmatrix} = \begin{bmatrix} \text{Transfer rate of} \\ \text{gas through the} \\ \text{packaging system} \end{bmatrix} + \begin{bmatrix} \text{Generation or} \\ \text{consumption rate of} \\ \text{gas from the product} \end{bmatrix}.$$

When the product is packaged will begin to consume O_2 from the packaging headspace and to generate CO_2 due its respiration, water vapor due its transpiration and ethylene. This will create a concentration differential between the inside and outside of the package resulting in an O_2 inlet and an outlet of the other gases to the external atmosphere. The composition of the different gases will be changing until a balance is reached between the metabolic processes in the product and the transfer of gases through the package. At this point, the MAP system becomes an equilibrium modified atmosphere packaging (EMAP), which is a kind of self-sustaining controlled atmosphere system that will be maintained until the product substrates are consumed or until storage conditions such as temperature are altered.

$$\begin{bmatrix} \text{Generation or} \\ \text{consumption rate of} \\ \text{gas from the product} \end{bmatrix} = \begin{bmatrix} \text{Transfer rate of} \\ \text{gas through the} \\ \text{packaging system} \end{bmatrix}.$$

Respiration Rate

Fresh produce shelf life depends on its respiration rate. Fruits such as mango and banana that have high respiration rates are highly perishable. The energy released during respiration is associated with the energy required to completely oxidize a mole of substrate to CO_2 and H_2O. In the case of a mole of hexose that is the most common carbohydrate, approximately 2880 kJ are released. Each hexose molecule is oxidized to 6 CO_2 molecules, using 32 molecules of adenosine diphosphate (ADP) and six molecules of O_2 to form 12 water molecules. Under normal physiological conditions, 50 to 60% of this energy is chemically captured to form 32 molecules of adenosine triphosphate (ATP), which are required for subsequent cell metabolic processes. As respiration is a series of oxidation reactions, the rate at which it is carried out will be related to the O_2 concentration in the medium, and when this concentration decreases, the energy released will also be reduced.

$$C_6F_{12}O_6(s) + 6O_2 \rightarrow 6CO_2 + 6H_2O + \text{energy}$$

In the case of carbohydrates such as glucose or fructose, when these are completely oxidized, the O_2 volume used per unit time will be practically the same as the volume of generated CO_2 per unit time. For organic acid oxidation, more CO_2 volume is produced than the used O_2 and for lipids oxidation, less CO_2 volume is generated than O_2 used. From the above, the respiratory quotient (RQ) is defined as the ratio of CO_2 produced to O_2 consumed per unit time. The RQ values range between 0.7 and 1.3 for the majority of fresh produce. High RQ values usually indicate anaerobic respiration in produce tissues or microbial activity that produces aldehydes and alcohols through fermentation. In such cases, a rapid change in the RQ can be used as an indication of the shift from aerobic to anaerobic respiration. Table shows estimated O_2 consumption, CO_2 production rates and RQ values for different fruits and vegetables.

Table: O_2 consumption and CO_2 production rates ($cm^3\ kg^{-1}\ d^{-1}$) and respiratory quotient at 20 °C and atmospheric air for different fresh produce.

Produce	r_{O2}	r_{CO2}	RQ
Apple	340–380	350–450	1.02–1.20
Avocado	3500–3800	2400–2700	0.70–0.72
Banana	3400–3600	2600–2700	0.75–0.76
Broccoli	2200–2400	2000–2150	0.88–0.90
Feijoa	4500–4800	3600–3800	0.79–0.80
Guava	850–1000	800–950	0.95–0.98
Strawberry	1050–1150	1000–1100	0.94–0.96
Tomato	640–680	700–730	1.05–1.07

Many mathematical models and equations have been used to represent the respiration process in fresh produce and the effect of the O_2 and CO_2 levels and storage temperature. These models include empirical equations, chemical kinetics and enzyme kinetics equations based on the Michaelis-Menten (MM) model. Among these equations stand out the Michaelis-Menten kinetics that have been successfully used for many products, specifically the equation of uncompetitive inhibition (MMU) wherein the inhibitor (CO_2 in this case) reacts with the enzyme-substrate complex, reducing the overall respiration rates at high concentrations. Nevertheless, the maximum O_2 consumption or CO_2 production rates are not influenced by the CO_2 levels in the atmosphere surrounding the product.

In the MMU kinetics, the rates of O_2 consumption and CO_2 generation are, respectively:

$$r_{O_2} = \frac{r_{O_{2max}} y_{O_2}}{K_{m_{O_2}} + y_{O_2}\left(1 + \frac{y_{CO_2}}{K_{mu_{CO_2}}}\right)}$$

$$r_{CO_2} = \frac{r_{O_{2max}} y_{O_2}}{K_{m_{CO_2}} + y_{O_2}\left(1 + \frac{y_{CO_2}}{K_{mu'_{CO_2}}}\right)}.$$

where r_{O2} and r_{CO2} are the O_2 consumption and CO_2 production rates, respectively (in $cm^3\ kg^{-1}\ d^{-1}$); r_{O_2max} and r_{CO_2max} are the maximum rates of O_2 consumption or CO_2 production, K_{mO_2} and K_{mCO_2} are the dissociation constants of the enzyme-substrate complex, and K_{muCO_2} and $K_{mu'CO_2}$ are the inhibition constants due to CO_2. In other cases, competitive, non-competitive or combined MM models have been used to represent the respiration rates.

Respiration rates are also directly influenced by temperature. As temperature increases, O_2 consumption and CO_2 production becomes higher, and this relationship has been satisfactorily described by considering each of the parameters of the Michaelis-Menten kinetics dependent on temperature by using the Arrhenius' Law.

Transpiration Rate

The transpiration rate can be defined as the water lost or evaporated in the product per unit time. The transpiration and water loss can be considered as a consequence of the heat transferred to and from the product and the vapor pressure difference between the product and the surrounding atmosphere. When the vapor pressure in the atmosphere is less than the vapor pressure in the product, it could be considered that both processes will contribute to transpiration. When the vapor pressures become equal, transpiration only will be seen as a consequence of the heat exchange to and from the product. This heat will be transferred to evaporate the product moisture mainly from the fraction of dissipated energy in the respiration process (that than is not used in the product metabolism), and from the temperature difference between the product and the surrounding atmosphere in the process of cooling. Some part of this energy transferred will change the product temperature as sensible heat, while another part will be converted to latent heat through moisture evaporation. If the system is in thermal equilibrium, as in a temperature-controlled chamber, the only heat that contributes to the moisture evaporation in the product will be the dissipated fraction of its respiratory heat. Thus, considering the above, the transpiration rate in a fresh produce can be expressed as follows:

$$r_{H_2O} = \frac{q}{\lambda}\left(\frac{RT}{PM_{H_2O}}\right) + k\,(a_{wp} - a_{wat}).$$

where r_{H_2O} is the transpiration rate (in $cm^3\ kg^{-1}\ d^{-1}$), q is the total heat transferred to the product ($kJ\ d^{-1}$), λ is the latent heat of moisture evaporation ($kJ\ kg^{-1}$), R is the gas constant, T is the temperature, P is the pressure, and M_{H_2O} is the water molar mass. In addition, k is total mass transfer coefficient due the water vapor differential (cm³ kg^{-1} d^{-1}), a_{wp} is the water activity in the product, and a_{wat} is the water activity in the surrounding atmosphere. The coefficient k will be influenced by temperature, and this relation can be expressed using the Arrhenius' Law. If the packing system is not in thermal equilibrium, it may be assumed that the effective heat contributing to the evaporation of water besides the water activity gradient is the dissipated fraction of respiratory heat to simplify the calculations in equation $r_{H_2O} = \frac{q}{\lambda}\left(\frac{RT}{PM_{H_2O}}\right) + k(a_{wp} - a_{wat})$. In that case, it is necessary to write the energy balance of the packaging system to estimate the temperature change in the product.

In case that the vapor pressure in the atmosphere (in the packaging headspace) becomes equal to the saturation pressure at the storage temperature, the water lost by the product as vapor will be condensed almost instantly inside the package. On the other hand, as a result of transpiration, the weight of the packaged product will be decreasing over the storage time which can bring undesirable effects on its quality. This weight loss can be quite significant when the differences in water activities between the product and the surrounding atmosphere are large or when the storage time is long, and for this reason, it should be considered in the modeling of the MAP system.

Gas Permeation through the Packaging Film

The permeability of a material to different gases will be determined by its molecular structure, thickness, surface area available for mass transfer, concentration gradient, pressure differential and temperature. For some biopolymers, the relative humidity should also be considered. When the film structure is uniform, without cracks or perforations, the gas will flow through the material by molecular diffusion. First, the gas will be adsorbed to the near face of the film, then the gas will diffuse through the film due to the concentration differential (from the point of greatest concentration to the point of lowest concentration), and thereafter, it will be desorbed from the farthest face from the film.

The permeation or transfer of a gas i through the material constituting the packaging system can be expressed as follows:

$$J_{fi} = -\frac{Q_i A(p_i - p_{i,out})}{L}\frac{P}{RT}.$$

where J_{fi} is the gas permeation through the continuous packaging film (mol d^{-1}), Q_i is the permeability coefficient of gas i in the packing material (cm³ mm m^{-2} atm^{-1} d^{-1}), and p_i and $p_{i,out}$ are the partial pressures of gas i in and out of the package (atm). P is the local pressure (atm), A is the film surface area (m²), and L is the thickness of the packaging material (mm). The permeability of the packaging films will depend on the storage temperature, and this can be represented by using the Arrhenius' Law considering the permeability coefficient Qi as a function of temperature.

Gas Transmission through Perforations in the Packaging System

The gas transfer through perforations made in each packaging film can be described by using the Fick's diffusion equation with a correction term (ε in equation below) due to the resistance to diffusion at the perforation. Thus, the transmission rate of gas i through the perforation (K_{Tri}, in $cm^3\ d^{-1}$) can be defined as follows:

$$K_{TRi} = \frac{D_i A_h}{L+\varepsilon}.$$

where D_i is the diffusion coefficient of the packaging film to gas i, Ah is the cross-sectional area of the perforations, L is the film thickness, and ε is the correction term due the diffusion resistance in the perforation channel. The correction term ε can be estimated as approximately 0.5 times the effective diameter (d_e) of the perforations, the effective diameter being related to the diameter of each perforation and to the total number of perforations through the expression $d_e = N^{1/2}d$.

The transmission rate of each gas through the perforation depends on the cross-sectional area and length (film thickness) of the perforation itself, the diffusivity of the gas in the air (gas mixture in the area to which the gas diffuses) and the concentration difference between the inside headspace and the external atmosphere. The dependence of the diffusion coefficients with respect to temperature can be described in the same way as the permeability coefficients using Arrhenius' law.

Balance Equations for the MAP System

In the headspace of a MAP system, the change in the O_2, CO_2, H_2O and N_2 concentrations can be described by using the mass balance equations for each gas considering the respiration and transpiration rates, the gas transfer through the packaging material, the gas transmission through the perforations in the package and the changes in volume or pressure. To write the equations, it is necessary to consider whether the packaging system is flexible or rigid.

If the packaging system is flexible (such as a bag), the packaging headspace will be varying with the change in the number of moles of gases in the system, while the pressure can be considered constant and equal to the atmospheric pressure. In this case, the balance equations will be as follows:

For O_2,

$$V\frac{dy_{O_2}}{dt} = K_{TRO_2}(y_{O_{2Out}} - y_{O_2}) + \frac{APQ_{O_2}}{L}(y_{O_{2Out}} - y_{O_2}) - r_{O_2}W - y_{O_2}\frac{dV}{dt}.$$

For CO_2,

$$V\frac{dy_{CO_2}}{dt} = K_{TRCO_2}(y_{CO_{2Out}} - y_{CO_2}) + \frac{APQ_{CO_2}}{L}(yCO_{2Out} - y_{CO_2}) + r_{CO_2}W - y_{CO_2}\frac{dV}{dt}.$$

For H_2O

$$V\frac{dyH_2O}{dt}=K_{TRH_2O}(y_{H_2O_{out}}-y_{H_2O})+\frac{APQ_{H_2O}}{L}(y_{H_2O_{out}}-y_{H_2O})+r_{H_2O}W-y_{H_2O}\frac{dV}{dt}.$$

For N_2,

$$V\frac{dy_{N_2}}{dt}=K_{TRN_2}(y_{N_{2Out}}-y_{N_2})+\frac{APQ_{N_2}}{L}(y_{N_{2Out}}-y_{N_2})-y_{N_2}\frac{dV}{dt},$$

where y_{O_2}, y_{CO_2}, y_{H_2O} and y_{N2} are the O_2, CO_2, water vapor and N_2 concentrations in the packaging headspace, respectively; y_{O_2out}, y_{CO_2out}, y_{H_2Oout} and y_{N_2out} are the gas external concentrations, V is the headspace volume (cm³), W is the product weight (kg), and t is the storage time (d).

The headspace volume change over time can be written as follows:

$$\frac{dV}{dt}=W(r_{CO_2}-r_{O_2}+r_{H_2O})+\sum_{i=1}^{4}\left((y_{iOut}-y_i)\left(K_{TRi}+\frac{APQ_i}{L}\right)\right),$$

where i is O_2, CO_2, water vapor or N_2.

If the headspace atmosphere is saturated (RH = 100 % and $y_{H_2O}=P_s/P$, where Ps is the saturation pressure), the change in the water vapor concentration with the storage time in the headspace will be zero, and the water evaporated in the product due to its transpiration (r_{H2O}) will be condensing inside the package. In this case, the equation $V\frac{dyH_2O}{dt}=K_{TRH_2O}(y_{H_2Oout}-y_{H_2O})+\frac{APQ_{H_2O}}{L}(y_{H_2O_{out}}-y_{H_2O})+r_{H_2O}W-y_{H_2O}\frac{dV}{dt}$ can be written as:

$$\frac{dy_{H_2O}}{dt}=0.$$

In this case, the contribution of r_{H_2O} to the volume change in the packaging headspace is not considered in equation $\frac{dV}{dt}=W(r_{CO_2}-r_{O_2}+r_{H_2O})+\sum_{i=1}^{4}\left((y_{iOut}-y_i)\left(K_{TRi}+\frac{APQ_i}{L}\right)\right)$.

If the packaging system is rigid, the packaging pressure will be varying with the change in the moles of gases in the system while the headspace volume can be considered constant along the storage time. In this case, the balance equations will be as follows:

For O_2,

$$V\frac{dy_{O_2}}{dt}=K_{TRO_2}\left((y_{O_2Out}-y_{O_2}\right)+\frac{APQ_{O_2}}{L}\left(y_{O_{2out}}-y_{O_2}\right)+\frac{N\pi d^4(p_{out}-P)y_{O_2}}{128\mu L_h}-r_{O_2}W-y_{O_2}\frac{V}{P}\frac{dP}{dt}.$$

For CO_2,

$$V\frac{dy_{CO_2}}{dt}=K_{TRO_2}\left(y_{CO_{2out}}-y_{CO_2}\right)+\frac{APQ_{CO_2}}{L}\left(y_{CO_{2out}}-y_{CO_2}\right)+\frac{N\pi d^4\left(p_{out}-P\right)y_{CO_2}}{128\mu L_h}-r_{CO_2}W-y_{CO_2}\frac{V}{P}\frac{dP}{dt}.$$

For H_2O

$$V\frac{dy_{H_2O}}{dt}=K_{TRH_2O}\left(\left(y_{H_2O_{out}}-y_{H_2O}\right)\right)+\frac{APQ_{H_2O}}{L}\left(y_{H_2O_{out}}-y_{H_2O}\right)+\frac{N\pi d^4\left(p_{out}-P\right)y_{H_2O}}{128\mu L_h}-r_{H_2O}W-y_{H_2O}\frac{V}{P}\frac{dP}{dt}.$$

For N_2,

$$V\frac{dy_{N_2}}{dt}=K_{TRN_2}\left(y_{N_{2out}}-y_{N_2}\right)+\frac{APQ_{N_2}}{L}\left(y_{N_{2out}}-y_{N_2}\right)+\frac{N\pi d^4\left(p_{out}-P\right)y_{N_2}}{128\mu L_h}-r_{H_2O}W-y_{H_2O}\frac{V}{P}\frac{dP}{dt},$$

where P_{out}-P represents the difference in pressure between the packaging headspace and the external atmosphere, which results in a gas flow through the packaging system that can be described by using the Poiseuille's Law. N is the number of perforations, d is the perforation diameter, μ is the viscosity of the gas mixture (which can be considered as air), and L_h is the effective length of the perforations (L + ε).

The headspace pressure change over time can be written as follows:

$$\frac{dP}{dt}=P\left(W\left(r_{CO_2}-r_{O_2}+r_{H_2O}\right)+\sum_{i=1}^{4}\left(\left(y_{iOut}-y_i\right)\left(K_{TRi}+\frac{APQ_i}{L}\right)+y_i\frac{N\pi d^4\left(P_{out}-P\right)}{128\mu L_h}\right)\right),$$

where i is again O_2, CO_2, water vapor or N_2.

As before, if the headspace atmosphere is saturated (RH = 100%), the change in the water vapor concentration with the storage time in the headspace will be zero, and the water evaporated in the product due to its transpiration will be condensing inside the package. In this case, the equation:

$$V\frac{dy_{H_2O}}{dt}=K_{TRH_2O}\left(\left(y_{H_2O_{out}}-y_{H_2O}\right)\right)\ldots r_{H_2O}W-y_{H_2O}\frac{V}{P}\frac{dP}{dt},$$

will be equal to equation $\frac{dy_{H_2O}}{dt}=0$. In this case again, the contribution of rH_2O to the pressure change in the packaging headspace is not considered in equation:

$$\frac{dP}{dt}=P\left(W\left(r_{CO_2}-r_{O_2}+r_{H_2O}\right)+\sum_{i=1}^{4}\left(\left(y_{iOut}-y_i\right)\left(K_{TRi}+\frac{APQ_i}{L}\right)+y_i\frac{N\pi d^4\left(P_{out}-P\right)}{128\mu L_h}\right)\right).$$

The weight change in the packed product over time can be determined as follows:

$$\frac{dW}{dt}=-Wr_{H_2O}\left(\frac{PM_{H_2O}}{RT}\right).$$

where W is the product weight, r_{H_2O} is the transpiration rate, P is the packaging pressure, R is the gas constant, T is the temperature, and M_{H_2O} is the water molar mass.

Model Considerations and Numerical Solution

By using the mass balance equations for the MAP system in combination with the equations to represent the product respiration and transpiration and the transfer of the gases through the package and perforations, it is possible to represent the change in the composition of the packaging headspace along the time under defined conditions. However, for the set of equations to be useful and achieve a numerical solution, it is required to establish several considerations and assumptions in the MAP model:

- During the packaging time, it can be assumed that the respiration and transpiration rates in the product are not affected by ripening or senescence considering that in general, at this time, the product is expected to remain in a kind of pseudoequilibrium state while consuming the substrates that allow the development of these processes.
- After reaching the thermal equilibrium in the packaging system, this is conserved throughout the storage.
- All temperature dependent processes can be modeled based on Arrhenius' law.
- The stratification of gases inside the package is negligible considering that at the sizes of packaging usually used in MAP, the concentration gradients are small due to the similarity in diffusivities of the gases involved and the small lengths of gas diffusion.
- The flexible packaging system acts as a kind of cylinder-piston when there is a change in the composition and the amount of gas in the packaging headspace, keeping the pressure approximately constant. Also in the rigid packaging system, the volume remains approximately constant throughout the storage time.
- The amount of moisture gained by the product is negligible with respect to the amount lost due to its transpiration.

The equations for the MAP system can be solved numerically by using a suitable program routine to integrate multiple differential equations. Whereas some components of the model solution, such as O_2 and CO_2 concentrations in headspace will change more rapidly than others such as water vapor concentration, headspace volume or product weight, the equations of the MAP system are characterized by being stiff or numerically unstable unless very small step sizes are used in the iteration process to find the solution. For this reason, it is recommended to use multistep implicit methods based on numerical differentiation formulas (NDF). When the packaging conditions such as concentration of gases and relative humidity in the external atmosphere, atmospheric pressure, product weight, packaging material, area and thickness of the flexible or rigid package and initial gas levels, headspace volume and pressure are defined, it is possible to estimate the change in the headspace atmosphere composition for a determined storage time.

Configuration of a Suitable MAP

Each MAP design must be optimized for a specific product, since agricultural products have different metabolisms to each other, and the MAP system must balance the processes of respiration, transpiration, ethylene production and gas permeation through the packaging that will be occurring simultaneously. Taking all of this into account, factors that affect both the metabolism of the packaged product and the gas permeation through the packaging should be considered for an optimal design of the MAP system.

Among these factors are firstly, the product maturity and ripeness stage, its weight and geometry and if it is whole or cut into fractions and secondly, the characteristics of the packaging system such as material, surface area, film thickness and headspace volume, and finally, the storage conditions such as temperature and external relative humidity. With regard to the packaged product, it is also necessary to know in advance its transpiration, respiration rates at the packaging conditions and the optimum concentrations of these gases and relative humidity that increase its shelf life. The design methodology of a successful MAP system can follow the steps shown in figure below:

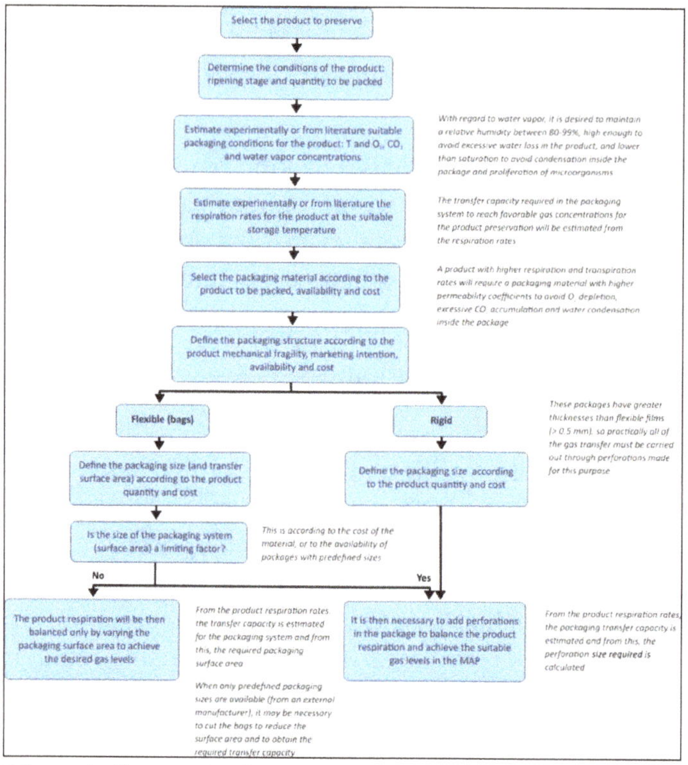

Schematic diagram for the configuration of a suitable MAP system.

Once the rates of the metabolic processes and characteristics of the packaged product are known, it is also necessary to determine the gas levels and temperature that are most favorable for its preservation. In this way, it is possible to determine the configuration of the packaging system in which the different processes are balanced, and the appropriate gas levels are reached steadily over the storage time.

First, it is necessary to define the product to be preserved and to determine its relevant characteristics such as maturity and ripeness stage, and configuration (whole or cut). It is then necessary to determine a suitable storage temperature that does not result in product deterioration (generally, it will be the most favorable temperature, but according to the specific conditions of distribution can be higher or even room temperature). Likewise, it is required to evaluate the levels of gases favorable for longer preserve the produce: O_2, CO_2 and HR. In order to determine these suitable conditions of temperature and gas concentrations, it may be required to perform experimental tests at various storage conditions or consult available data.

To determine the gas transfer capacity required in the MAP system, it is necessary to pre-estimate the respiration and transpiration rates of the product to be packed. These rates can be determined as a function of gas levels and temperature by using equations:

$$r_{O_2} = \frac{r_{O_{2max}} y_{O_2}}{K_{m_{O_2}} + y_{O_2}\left(1 + \frac{y_{CO_2}}{K_{mu_{CO_2}}}\right)}, \quad r_{CO_2} = \frac{r_{O_{2max}} y_{O_2}}{K_{m_{CO_2}} + y_{O_2}\left(1 + \frac{y_{CO_2}}{K_{mu'_{CO_2}}}\right)},$$

$$r_{H_2O} = \frac{q}{\lambda}\left(\frac{RT}{PM_{H_2O}}\right) + k\,(a_{wp} - a_{wat}),$$

and knowing previously the parameters of each equation that must be obtained experimentally or from other studies.

Once the aspects corresponding to the product to be preserved are known, then the configuration of the packaging system for its preservation is followed. First, it is necessary to evaluate of the packaging material to be used and estimation of its permeation rates to the different gases. This evaluation will of course depend on the transfer capacity required to properly balance the metabolism of the product. For a product with high respiration rates, a packing material with high permeability to O_2 and CO_2 is required. Likewise, for a product with high transpiration rate, a packing material with a high permeability to water vapor will be necessary. Other considerations must be made such as the availability and cost of the packaging material. Many biodegradable polymers such as PLA or EVOH are suitable for MAP systems but their availability on the market is still limited, and their cost is still high compared to synthetic materials such as PP and LDPE.

To determine the required transfer capacity in the package to reach the suitable gas levels (O_2, CO_2 or RH), it is necessary to estimate the transfer area of the packaging material and/or the size of the perforations required by using the mass balance equations of the MAP system in steady state. If the gas transfer capacity in the package is enough to compensate the produce respiration and transpiration, after a certain time an atmosphere of equilibrium will be reached inside the MAP system and for a flexible packaging, for example, equations $V\frac{dy_{O_2}}{dt}=K_{TRO_2}(y_{O_{2Out}}-y_{O_2})...W-y_{O_2}\frac{dV}{dt}$,

$V\frac{dy_{CO_2}}{dt}=K_{TRCO_2}...r_{CO_2}W-y_{CO_2}\frac{dV}{dt}$, $V\frac{dyH_2O}{dt}=K_{TRH_2O}(y_{H_2O_{out}}-y_{H_2O})...r_{H_2O}W-y_{H_2O}\frac{dV}{dt}$

and $V\frac{dy_{N_2}}{dt}=K_{TRN_2}(y_{N_{2Out}}-y_{N_2})...-y_{N_2}\frac{dV}{dt}$ may be rewritten as follows:

$$\left(K_{TRO_2}+\frac{APQ_{O_2}}{L}\right)\left(y_{O_{2out}}-y_{O_{2eq}}\right)-r_{O_2}W-y_{O_{2eq}}\frac{dV}{dt}=0$$

$$\left(K_{TRCO_2}+\frac{APQ_{CO2}}{L}\right)\left(y_{CO_{2\,Out}}-y_{CO_{2\,eq}}\right)+r_{CO_2}W-y_{CO_{2\,eq}}\frac{dV}{dt}=0$$

$$\left(K_{TRH_2O}+\frac{APQ_{H_2O}}{L}\right)\left(y_{H_2O_{Out}}-y_{H_2O_{eq}}\right)+r_{H_2O}W-y_{H_2O\,eq}\frac{dV}{dt}=0$$

where $y_{O_2\,eq}$, $y_{CO_2\,eq}$ and $y_{HO_2\,eq}$ are the equilibrium O_2, CO_2 and water vapor concentrations. The change in the free package volume over time dv/dt, is substituted from equation $\frac{dV}{dt}=W(r_{CO_2}-r_{O_2}+r_{H_2O})+\sum_{i=1}^{4}\left(\left(y_{iOut}-y_i\right)\left(K_{TRi}+\frac{APQ_i}{L}\right)\right)$. The N_2 balance is not necessary since N_2 concentration (molar fraction) is a function of the O_2, CO_2 and water vapor concentrations considering that the sum of all concentrations must be equal to 1, and that the presence of other gases is negligible. In the case of rigid packages, equations $\frac{dy_{H_2O}}{dt}=0$, $V\frac{dy_{O_2}}{dt}=K_{TRO_2}\left(\left(y_{O_2Out}-y_{O_2}\right)...r_{O_2}W-y_{O_2}\frac{V}{P}\frac{dP}{dt}\right.$,

$$V\frac{dy_{CO_2}}{dt}=K_{TRO_2}\left(y_{CO_{2out}}-y_{CO_2}\right)...r_{CO_2}W-y_{CO_2}\frac{V}{P}\frac{dP}{dt}$$

and

$$V\frac{dy_{H_2O}}{dt}=K_{TRH_2O}\left(\left(y_{H_2O_{out}}-y_{H_2O}\right)\right)...r_{H_2O}W-y_{H_2O}\frac{V}{P}\frac{dP}{dt}$$

can be rewritten similar to equations:

$$\frac{dW}{dt}=-Wr_{H_2O}\left(\frac{PM_{H_2O}}{RT}\right),\ \left(K_{TRO_2}+\frac{APQ_{O_2}}{L}\right)\left(y_{O_{2out}}-y_{O_{2eq}}\right)-r_{O_2}W-y_{O_{2eq}}\frac{dV}{dt}=0,$$

$$\left(K_{TRCO_2} + \frac{APQ_{CO2}}{L}\right)\left(y_{CO_2\,Out} - y_{CO_2\,eq}\right) + r_{CO_2} W - y_{CO_2\,eq}\frac{dV}{dt} = 0,$$

and $\left(K_{TRH_2O} + \frac{APQ_{H_2O}}{L}\right)\left(y_{H_2O_{Out}} - y_{H_2O_{eq}}\right) + r_{H_2O} W - y_{H_2O\,eq}\frac{dV}{dt} = 0$ with the change in concentration of each gas with respect to time equal to zero in steady state. In that case, the change in the headspace pressure over time dP/dt is substituted from equation:

$$\frac{dP}{dt} = P\left(W\left(r_{CO_2} - r_{O_2} + r_{H_2O}\right) + \sum_{i=1}^{4}\left(\left(y_{iOut} - y_i\right)\left(K_{TRi} + \frac{APQ_i}{L}\right) + y_i\frac{N\pi d^4\left(P_{out} - P\right)}{128\mu L_h}\right)\right).$$

The meaning of equations $\left(K_{TRO_2} + \frac{APQ_{O_2}}{L}\right)\left(y_{O_{2out}} - y_{O_{2eq}}\right) - r_{O_2} W - y_{O_{2eq}}\frac{dV}{dt} = 0,$

$$\left(K_{TRCO_2} + \frac{APQ_{CO2}}{L}\right)\left(y_{CO_2\,Out} - y_{CO_2\,eq}\right) + r_{CO_2} W - y_{CO_2\,eq}\frac{dV}{dt} = 0,$$

and,

$$\left(K_{TRH_2O} + \frac{APQ_{H_2O}}{L}\right)\left(y_{H_2O_{Out}} - y_{H_2O_{eq}}\right) + r_{H_2O} W - y_{H_2O\,eq}\frac{dV}{dt} = 0,$$

is that to reach a determined steady concentration of gas, a given transfer capacity by the package will be required to match the produce respiration and transpiration. To achieve this transfer capacity, the type of packaging material, the package surface area and the number and size of perforations can be modified. By knowing the produce respiration and transpiration rates, it is possible define a packaging system to achieve a self-sustaining controlled atmosphere with favorable gas levels.

Depending whether the size of the package is defined or not, two possible configurations arise. In the first configuration, if the size is not defined, it is possible to achieve the required transfer capacity of the package simply by increasing the film surface area. In this case, the K_{TR} values of each gas are zero in the absence of perforations and four variables will be taken: The equilibrium O_2, CO_2 and water vapor concentrations, y_{O_2eq}, y_{CO_2eq}, y_{H_2Oeq} and the required surface area. As there are four variables and three independent equations, fixing one of the former, zero degrees of freedom are left, and the remaining three variables are determined in the calculation. What this means is that it is only possible to configure the system to specifically set only a gas concentration, that of O_2, CO_2 or water vapor. If it is wanted to set the equilibrium O_2 concentration in the MAP system, for example, the equilibrium CO_2 and water vapor concentrations (and of course the surface area) will be already defined and cannot be set. As is not possible to isolate A and $y_{O2\,eq}$ in equations $\left(K_{TRO_2} + \frac{APQ_{O_2}}{L}\right)\left(y_{O_{2out}} - y_{O_{2eq}}\right) - r_{O_2} W - y_{O_{2eq}}\frac{dV}{dt} = 0,$

$$\left(K_{TRCO_2} + \frac{APQ_{CO2}}{L}\right)\left(y_{CO_2\,Out} - y_{CO_2\,eq}\right) + r_{CO_2}W - y_{CO_2\,eq}\frac{dV}{dt} = 0,$$

and,

$$\left(K_{TRH_2O} + \frac{APQ_{H_2O}}{L}\right)\left(y_{H_2O_{Out}} - y_{H_2O_{eq}}\right) + r_{H_2O}W - y_{H_2O\,eq}\frac{dV}{dt} = 0,$$

it is necessary to calculate the value of the former (and the other concentrations) iteratively.

It is worth noting that the water vapor concentration in the packaging headspace will be about 1% (mole fraction) in almost all situations due to its low saturation pressure with respect to the atmospheric pressure at the typical storage conditions (e.g., at 10 °C and 1 atm, 1.21% water vapor is equivalent to 100% RH). Considering this, it is possible to somehow simplify calculations to set a defined O_2 concentration and find the required area by assuming the water vapor concentration (between 1 and 2% mole fraction) and run iterations using only two equations instead of three.

In the second configuration, if the size of the package is defined, the surface area will be constant and to achieve the required transfer capacity is necessary to make perforations. In this case, the equilibrium gas concentrations, y_{O_2eq}, y_{CO_2eq}, y_{H_2Oeq} and the transmission coefficients through the perforation, K_{TRO_2}, K_{TRCO_2}, K_{TRH_2O} and K_{TRN_2} will be taken as variables. However, from equation $K_{TRi} = \frac{D_i A_h}{L + \varepsilon}$, it can be seen that just as the diffusion coefficients of each gas in the atmosphere, the thickness of the bag and the ε correction term are already defined, the real variables are the diameter and the number of perforations in the package. If the number of perforations is fixed, only the diameter is variable, and if the perforation diameter is fixed, only the number of these may be varied, although it is better to define the diameter to estimate an integer number of perforations. Four variables will then be taken: the equilibrium O_2, CO_2 and water vapor concentrations, y_{O2eq}, y_{CO2eq}, y_{H2Oeq} and the diameter or number of perforations. As in the first setting, when one of the concentrations is fixed, zero degrees of freedom are left, and the remaining concentration and the diameter or the number of perforations are determined in the iterative calculation.

In general, for n independent components in the packaging headspace, it will be n + 1 variables (counting the variable related to the transfer capacity in the package, e.g., the diameter of perforations), and n independent equations. By setting one of the gas concentrations, the system can be solved.

In the case of rigid packages where a higher transfer capacity is given by the perforations, the same procedure explained for the second configuration can be followed.

Depending on the configuration defined for the MAP system and the initial gas concentrations in the headspace, it can take more or less time to reach a steady state and obtain the desired gas concentration. It can then be quite convenient to seal the package with an initial concentration close to the desired one to achieve equilibrium more quickly.

Permissions

All chapters in this book are published with permission under the Creative Commons Attribution Share Alike License or equivalent. Every chapter published in this book has been scrutinized by our experts. Their significance has been extensively debated. The topics covered herein carry significant information for a comprehensive understanding. They may even be implemented as practical applications or may be referred to as a beginning point for further studies.

We would like to thank the editorial team for lending their expertise to make the book truly unique. They have played a crucial role in the development of this book. Without their invaluable contributions this book wouldn't have been possible. They have made vital efforts to compile up to date information on the varied aspects of this subject to make this book a valuable addition to the collection of many professionals and students.

This book was conceptualized with the vision of imparting up-to-date and integrated information in this field. To ensure the same, a matchless editorial board was set up. Every individual on the board went through rigorous rounds of assessment to prove their worth. After which they invested a large part of their time researching and compiling the most relevant data for our readers.

The editorial board has been involved in producing this book since its inception. They have spent rigorous hours researching and exploring the diverse topics which have resulted in the successful publishing of this book. They have passed on their knowledge of decades through this book. To expedite this challenging task, the publisher supported the team at every step. A small team of assistant editors was also appointed to further simplify the editing procedure and attain best results for the readers.

Apart from the editorial board, the designing team has also invested a significant amount of their time in understanding the subject and creating the most relevant covers. They scrutinized every image to scout for the most suitable representation of the subject and create an appropriate cover for the book.

The publishing team has been an ardent support to the editorial, designing and production team. Their endless efforts to recruit the best for this project, has resulted in the accomplishment of this book. They are a veteran in the field of academics and their pool of knowledge is as vast as their experience in printing. Their expertise and guidance has proved useful at every step. Their uncompromising quality standards have made this book an exceptional effort. Their encouragement from time to time has been an inspiration for everyone.

The publisher and the editorial board hope that this book will prove to be a valuable piece of knowledge for students, practitioners and scholars across the globe.

Index

www.ingramcontent.com/pod-product-compliance
Lightning Source LLC
LaVergne TN
LVHW080617210926
848485LV00003B/174
9781639873357